Modern Manufacturing Processes

Modern Manufacturing Processes

James Brown

Industrial Press Inc.

Library of Congress Cataloging-in-Publication Data

Brown, James A.
Modern manufacturing processes / James Brown.
240 p. 11.7 x 17.8 cm.

ISBN 0-8311-3034-2
1. Manufacturing processes.   I. Title.
TS183.B76 1991   90-48046
670.42 — dc20   CIP

Industrial Press Inc.
200 Madison Avenue
New York, New York 10016-4078

First Printing

First Edition

Modern Manufacturing Processes

Composition by V&M Graphics, Inc., New York, New York.
Printed and bound by Quinn Woodbine, Woodbine, New Jersey.

2   4   6   8   9   7   5   3

CONTENTS

PREFACE

There is a significant reason for manufacturing professionals to read and study this book: all professionals should be concerned with product cost. The military wants "a bigger bang for the buck." Industry wants foreign markets. None of this is possible until we learn to manufacture products at lower cost.

Usually the designer is the individual most involved with determining product cost because he draws the lines which indicate the most reasonable manufacturing process. But the engineers who sign off on his drawings are also responsible, and so is the program manager, and even the procurement department which places orders for part manufacture and component purchase. If these involved people could take a second look at the design and the quantities required, perhaps small modifications would allow a different, less costly manufacturing process. Knowledge and understanding of the contents of this book should facilitate economical manufacture.

Chapters in this book are placed in a sequence which promotes continuity, especially if the reader has a minimum knowledge of the subject.

The first two chapters, "Metal Injection Molding" and "Aluminum Brazing," set the pace for the book. The first because the technology is fairly new and is not widely known, and the second because it describes an old process filling an urgent need in modern productivity. The next five chapters deal with various metal surface treatments. Chapters 8–11 cover unusual and unrelated subjects. Chapters 12–18 are concerned with metal working processes which are not used often enough. Chapters 19–21 are on mass production machine deburring. Chapter 22 is similar to 21 in that both describe electrochemical processes. The remaining six chapters are important to complete the manufacturing engineer's knowledge of his "trade."

They include two chapters on materials which will be increasingly significant as we approach the 21st century.

Although many of the processes described in this book are well established, many people responsible for product cost do not fully understand their value. The author does not claim to provide 100 percent coverage of the subjects. He has, however, tried to give sufficient information to guide those anxious to "find a better way." If some individuals obtain assistance from the material included in this book, the effort of writing it will have been worthwhile.

ACKNOWLEDGEMENTS

Acknowledgements and grateful thanks are due to the following people and companies for information supplied and for their kind permission to publish data, photographs, and illustrations. Their cooperation is appreciated because the book could not have been written without it.

Chapter	Contributors
1. Metal Injection Molding	Prof. R. M. German — Rensselaer Polytechnic Institute, Troy, NY; Dr. A. Nyce — Gorham Advanced Mat. Institute, Gorham, ME
2. Aluminum Brazing	The Aluminum Assoc., Washington, DC
3. Hardcoating	Duralectra Metal Finishing Co., Natick, MA; Durafilm Corp., Holliston, MA
4. Titanium Nitride Coating for Superior Tool Performance	Balzers Tool Coating Co., North Tonawanda, NY
5. Poly-Ond	Poly Plating Inc., Chicopee, MA
6. Electrolizing	Electrolizing Inc., Providence, RI
7. Powder Coating	Plastonics Inc., Hartford, CT
8. Electroforming	Gar Electroforming Co., Danbury, CT
9. Electrical Discharge Machining	Elox Corp., Davidson, NC; N.E.T.W. Inc., New Britain, CT; Easco-Sparcatron Inc., Whitmore Lake, MI

10. Abrasive Waterjet Cutting	Mr. A. Hitchcock and *The Advanced Materials & Processes Magazine*
11. Magneform	Maxwell Laboratories, San Diego, CA
12. Fineblanking	Connecticut Fineblanking Corp., Shelton, CT
13. Cold Forming — Coining	D. Hunter, Cousino Metal Products Inc., Toledo, OH; Society of Manufacturing Engineers, Dearborn, MI
14. Hydroforming	Roland Teiner Co., Everett, MA
15. Metal Spinning	Metal Spinning Division of American Metal Stamping Assoc., Richmond Heights, OH
16. Shearforming	Floturn Inc., Cincinnati, OH; Autospin Inc., Carson, CA
17. Orbital Cold Forging	Schmid USA Holding, Inc., Goodrich, MI
19. Thermal Energy Deburring	Surftran Corp., Madison Heights, MI
20. Abrasive Flow Deburring	Extrude Hone Corp., Irwin, PA
21. Electrochemical Deburring	Anocut Inc., Elk Grove Village, IL
22. Electrochemical Machining	Anocut Inc.; *Machine & Tool Blue Book Magazine*, Carol Stream, IL
23. Computer Integrated Manufacturing	Valisys Corp., Santa Clara, CA
24. Advanced Composites	Textron Specialty Materials, Lowell, MA; *Manufacturing Engineering Magazine*, Dearborn, MI; *Machine Design Magazine*, Cleveland, OH
25. Ultrasonic Technology	Dukane Corp., St. Charles, IL; Bullen Ultrasonics Inc., Eaton, OH
26. Hot Isostatic Pressing	Industrial Materials Technology Inc., Andover, MA; Howmet Corp., Whitehall, MI

1

METAL INJECTION MOLDING

PROCESS OVERVIEW

Metal injection molding (MIM) is the process of mixing elemental or alloyed powders with thermoplastic binders. The binders are selected to deliver the optimum flow characteristics to facilitate uniform distribution of the material in the mold.

Under moderate pressures and temperatures, the material is extruded into a mold. The material hardens as it cools and is removed from the mold in what is called the "green" condition. At this point, the part is larger than the desired shape by the amount of binders used in the mixture. These binders are preferentially and selectively extracted from the "green" part. Generally, a low-temperature furnace treatment slowly evaporates the binder over a 1–3 day period. The parts are then sintered in a high-temperature furnace under controlled atmospheric conditions, and a temperature profile is selected to remove the remaining binders and present a finished part with the desired physical properties. Obviously, the entire process lends itself admirably to microprocessor control.

PROCESS DETAILS

Manufacturing is a dynamic institution. New means of producing parts are continually surfacing. Metal injection molding is one of them. This process provides a new freedom in part geometry which eliminates the restrictions of the more conventional metal working processes; sometimes two or three parts can be molded as one. Although MIM had progressed for about 15 years in the laboratory, during the 1980's it entered the arena of competitive production. The point is that, although this is a newly emerging metal working technique, it has endured a long, arduous development.

The MIM process makes intricate parts from a mixture of finely divided metal powders (5–10 microns in size) and a binding agent which could be one of several thermoplastics. The mixture ordinarily contains 10–35% binders. Since the binders are later removed by either heat, solvents, or both, the molds should be larger to accommodate the 10–35% shrinkage which follows. As could be expected, this is an area that is proprietary and contains a closely guarded secret for each vendor. There are at least four companies which license their processes.

Dr. Ray Weich, of Multi-Material Molding in San Diego, CA, holds three patents on the use of thermoplastic binders. One patent uses a solvent extraction process to remove binders prior to sintering. Several manufacturers have been licensed to use these patents. Multi-Material Molding has also introduced a continuous binder removal process.

New Industrial Techniques Inc., Coral Springs, FL, is using the Rivers Process, licensed by Hayes International, Kokomo, IN. (Some people identify this as the Cabot Process.) This process uses water soluble methylcellulose as the binder. A 0.25″-thick part could be debinderized in less than 4 hours, and could possibly enable a full cycle—molding, debinderizing, and sintering —to take place in one day (one shift).

Two of the MIM process features which create expense are the length of time to debinderize and to sinter. If this 4–5 day cycle could be cut to 1 day, much expense would be avoided. That is why so much experimentation is going on in this area.

There is another process which removes binder with a solvent. This permits debinderization for a ½″ section in 4 hours. Most binderization processes in current use are based on the above processes customized for the individual vendor.

Recently, a list of 14 different binder formulations was published. Each binder has advantages and disadvantages, and most vendors experiment to enhance the former and eliminate the latter. There have been problems in these attempts to accelerate debinderization. Several problems have been solved, but some remain. Because of this, there seem to be specific areas where each type of binder excels. At least one MIM vendor uses more than one basic type of binder for this reason. But experimentation continues just as it does in most of the conventional processing methods. The latest work in binder systems tends to raise the debinderization temperature as high as 1250°F.

PROCESS DESCRIPTION

MIM is a recently developed technique which is applicable to small, complex, steel or ceramic parts. There are special machines used to accelerate and efficiently complete each step. Each part processed must undergo four steps: mixing, molding, debinderizing, and sintering.

MIXING: Metal powders are mixed with binders in a small mechanized bucket like a cement mixer. The material is pelletized and poured into the injection molding machine hopper.

MOLDING: This is done in a standard plastic injection molding machine just as if a glass-filled nylon part were being made. The feedstock is heated, then injected into a mold. It cools and solidifies in the same way the nylon part would. Then it is ejected as a "green" part. A simple single-cavity mold may cost $4000. But if production quantities warrant a higher mold cost, a multiple-cavity mold, 2–12 cavities (or more), could be utilized; for example, $20,000 spent for a multicavity mold could lower a piece price from $3.00 to $1.50. If volume was 100,000 per year, it would not take long to amortize the additional tooling cost. The parts are loaded onto large ceramic shelves, each shelf holding hundreds of parts. The shelves, holding "green" parts, are placed in the debinderizer.

DEBINDERIZING: This is done in a low-temperature furnace which seldom goes higher than 400°F. If more than one plastic makes up the binder, the temperature will sequentially evaporate the various constituents. The parts (now called "brown") should by now have shrunk down to nearly the final size. The list of temperatures used and the soak times required make this work ideal for computer control. The ceramic shelves, loaded with parts, are now placed in the sintering furnace.

SINTERING: Here the parts attain the condition known as "white." They assume their final critical shape and properties. This is another area which should be computer controlled to maintain proper atmospheric conditions with a temperature profile designed to extract any remaining binders and produce the required density, mechanical properties, and surface finish. In this step of the process, the final temperature is close to the melting point of the steel material. This diffuses the powder and increases density to about 94–97% of wrought material. The shrinkage is uniform along the X, Y, and Z axes and, as a result, critical dimensions and complex shapes can be attained.

SECONDARY OPERATIONS

If the parts could be produced without secondary operations, it would be most economical. A good MIM mold designer knows what to expect when he views the part drawing. It is always a good idea to discuss the part configuration with him before the drawing is finalized. This is true whether the part is MIM, sheet metal, a casting, or a forging. But if some undesirable design characteristic cannot be changed, secondary operations should be considered. Internal threading can be facilitated by positioning the hole and countersinking it in the MIM process. This leaves only a tapping operation without requiring deburring. In some cases, both internal and external threads can be economically molded.

Normally, $\pm 0.002''$ tolerance can be held in a $1''$ length. But if a tighter tolerance is specified, make the part slightly oversized and then machine that surface. Some parts will require coining or sizing to maintain a consistent quality. During the sintering step, cavities may be slightly egg shaped or tapered instead of round.

Most MIM parts require some type of finishing operation. Simple deburring or breaking sharp edges can usually be accomplished by tumbling in rotating tubs. The final use of the part, and the material of the part, dictate what else will be done.

Usually, there is no need to passivate the stainless steels. Some ferrous parts may be treated with a corrosion preventative oil. Most conventional surface finishes may be used, such as nickel or chrome plating.

Sometimes heat treatment can be done immediately after parts are removed from the sintering furnace. At this time, the parts are in the annealed condition. Some can be hardened simply by quenching; other materials can be nitrided or carburized. Welding or brazing may also be performed.

DESIGN CRITERIA

As mentioned earlier, it makes good sense to confer with the MIM mold designer before going into production. But often modifications to part design cannot be accommodated. Let the MIM vendor quote your job anyway. There are many steps he can take to meet your specifications, even the difficult ones. He understands the process so well that he will be able to hold a few tight tolerances one way or another. When his first samples deviate from design dimensions, he will modify the tooling.

It has been said that any part smaller than a golf ball might be a candidate for MIM. Another way to view size is to try to stay within $1'' \times 1'' \times \frac{1}{2}''$. However, larger pieces have been processed. One manufacturer showed parts $\frac{3}{4}''$ thick $\times$ 2'' long. Theoretically, there is no limit to size, you are limited only by your imagination.

Any geometry which can be achieved by plastic injection molding can also be produced by MIM in steel or ceramic. Visualize a glass-filled nylon part — if the shape can be made in nylon, it can also be done by MIM. If the design requires a minimum radius, it could actually be made sharp. Holes and undercuts can also be handled the same way you would do it in nylon.

In a molding process, parting lines cannot be avoided. Wherever the mold halves come together, there will be some extrusion of material. By designing a small flat at this parting line, the flash will not affect the part function.

In all types of molding or casting, good design practice avoids large differences in material cross section and long thin areas. General tolerances in MIM are the same as for plastic injection molding:

Angular = 0.5 degrees

Linear = 0.003'' per inch of length.

It may be possible to hold one dimension closer, but for consistency, a machine finish may be required.

The following are some parts which have been made by MIM:

stainless steel orthodontic braces,

tungsten carbide grinding burrs,

kovar electronic packages,

inconel and aluminum oxide gears and spray nozzles,

nickel steel firearm parts,

high-speed printer hammers,

automobile parts and consumer products,

niobium rocket thrust chambers (this niobium part is 6'' long and works in a 2500°F environment).

Production equipment is sometimes designed or modified for specific alloys; 300 and 400 series stainless steels are being successfully produced by MIM. Nickel iron seems to be the most popular (easiest to handle) alloy in current use. AISI 4340 and tool steels are now being processed where high strength and hardness requirements predominate. A 3% silicon steel is available for magnetic applications.

The alloy capabilities of MIM seem endless. Apparently, vendors are willing to try any alloy you may need, providing the metal powder is available; but not all metals are available in powder form. However, any powders manufactured can be mixed in any desired combinations.

Most MIM vendors will show tables and charts comparing physical data such as densities, chemical composition, tensile and yield strengths, elongation, and hardness and magnetic qualities of materials used in MIM production. These will be provided to guide your material selection. Reading these tables is a sensible first step. However, for the sake of quality control, you must subject the first MIM samples you receive to the laboratory tests which concern you. As in any production, especially those not dealing with wrought material, it is possible to obtain an inferior product. In MIM work, that simply means that some step in the process needs adjustment, and this is the time and place to learn that.

In the early days of MIM open competition, one company had some samples made. It would have meant a change in manufacturing methods if successful. It would also have reduced part cost from $32 to $2 each. The samples looked beautiful — just what the doctor ordered. All dimensions were satisfactory. Nevertheless, some parts were sent to the laboratory for physical tests. The project engineer was surprised to learn that there was a large scatter in the value of elongation, ultimate, and yield strengths. The specimen's break surfaces indicated the existence of a sintering problem. In this case, a modification of the sintering step corrected all ills. But without the tests, this correction would not have been made.

ECONOMICS

MIM favors small, intricate sections. In fact, when small parts have to be made, this technology has a cost advantage over some conventional processes such as powder metallurgy, investment casting, and plain machining. It conserves time and material, and the mold material trimmings can be ground and recycled.

MIM material powders cost much more than powder metallurgy powders (PM) because they are finer in size, and because the quantity made and sold is so small in comparison. Ceramic powders are available much finer than the 5 microinch steel powders. Because of these fine powders (1 microinch), mirror-like surfaces are being produced much more cheaply than before. Ferrous parts with parallel faces can be made by punch press or conventional PM more cheaply than by MIM. Complex steel shapes, however, are natural candidates for MIM.

A local defense contractor had 24 parts cost estimated by an MIM vendor. None of the tooling exceeded $20,000 and most of them were multicavity molds. The piece part prices averaged 100–600% less by MIM than by any other method. (Naturally, the tool cost rises when you go for multicavity molds, but the piece part cost falls. This is where the economies of mass production must be compared to the cost of tooling so an intelligent decision can be made.) One of those 24 parts required a quantity of only 1000 pieces. But the tooling cost was easily amortized and money was saved by eliminating difficult machining. Tooling costs and their design are comparable to those of plastic injection molding. According to some estimates, MIM sales in 1989 exceeded $20,000,000.

HEAT TREATMENT

Ferrous MIM parts of medium or high carbon content (0.3% or more) may be quenched directly from the sintering furnace for improved wear resistance and strength. The various other alloys present determine hardenability and, perhaps, a preferred method of heat treatment. An oil quench is often recommended because oil is not corrosive if retained. If the steel parts have a low carbon content, most could easily be carburized.

MIM MATERIALS

Most metals or ceramics available in powder form smaller than 10 microns can be used in this process. Many MIM vendors have experience with the following materials:

1. iron nickel alloys, 2–50% nickel;
2. stainless steels, 300–400 series and others;
3. pure metals: iron, nickel, molybdenum, cobalt, tungsten, columbium, niobium;
4. high strength steel alloys, 4130–4340;
5. a variety of tool steels;
6. kovar, invar, inconel;
7. tungsten carbide alloys;
8. aluminum oxide (99.9%);
9. zirconia oxide;
10. silicon carbide;
11. silicon nitride; and
12. ceramic and ceramic–metal composites.

EQUIPMENT FOR MIM OPERATIONS

In a production facility, the following equipment is generally found.

Raw Material Treatment:
   ball mills,
   mixers,
   material granulizers,
   material dryers,
   precision scales.

Molding Department:
   injection molding machines,
   material granulizers,
   trimming and handling equipment,
   precision scales,
   microscopes.

Processing and Sintering Department:
   batch processor,
   debinderizing furnace,
   sintering furnace,
   ceramic kilns,

microscopes,
precision scales,
handling equipment.

Quality Control Department:
comparators,
microscopes,
precision scales,
density tester,
carbon analyzer,
tensile tester,
computer,
measurement equipment.

Development Laboratory and Material Testing:
thermographic analysis,
material mixer,
centrifuge,
ball mill,
high-temperature vacuum atmospheric sintering furnace,
sedimentation measuring device,
debinderizer,
metallurgical sample preparation equipment,
bench metallograph,
carbon analyzer,
scanning electron microscope,
precision scales,
viscometer,
dewpointer,
gas analyzer.

GENERAL OBSERVATIONS

"Injectalloy" is the Remington Arms Co., Inc.'s name for their MIM process. Similarly, "Injectomet" is the Engineered Sinterings and Plastics Company name for their process. These two companies, plus thirty-one others, are American fabricators using MIM.

It is a good idea to work with more than one MIM vendor. Most have fairly small operations, and their facilities would be swamped by a few large orders. Consequently, to safeguard your schedule, it will be necessary to investigate several vendors.

To avoid disagreements over the delivery of quality parts, it is also a good idea to make the MIM vendor responsible for the completed parts, including all secondary machining, heat treatment, and surface finishing.

Each step of the MIM process is constantly being evaluated for improvement possibilities. Originally, the debinderizing and sintering steps caused about 75% of the MIM process expense. Now this percentage has dropped dramatically and will approach 30% in some cases. Of course, this depends on the powders used as some are very expensive.

Currently, there are about thirty-three vendors in the U.S.A. available to do MIM work for you. A list can be obtained from the Metal Powder Industries Federation in Princeton, NJ. The interest in MIM is worldwide, and meetings of the Federation are attended by people from England, France, Germany, Italy, Holland, Japan, Canada, India, Taiwan, Poland, Denmark, and Finland.

2

ALUMINUM BRAZING

GENERAL INFORMATION

If you bring two pieces of metal within 4 angstroms ($\mathring{A} = 10^{-10}$ meter) of each other, interatomic attraction will bind them together permanently. This phenomenon is the basis of brazing and soldering, and is accomplished by "wetting" the metals to be joined with molten metal which forms the joint when it cools.

The American Welding Society has simple definitions for brazing and soldering. Above 800°F, the process is called brazing and uses brazing filler metal; below 800°F, the process is called soldering and the molten metal is called solder. Welding differs from these two processes because the base metals to be joined are themselves molten at the moment of joining.

There are many advantages to brazing.

Strong, uniform, leakproof joints can be made inexpensively and rapidly by modern techniques. Inaccessible joints and parts which may not be joinable by any other method can often be joined by brazing.

Complicated assemblies with thick and thin sections, odd shapes, and differing wrought and cast aluminum can be made into one integral component by a single trip through a brazing furnace or a dip pot. Metal as thin as 0.006" and as thick as 6" can be brazed.

Brazed joint strength is very high. The nature of the interatomic bond is such that even a simple joint, when properly designed, will have strength equal to or greater than the base metal.

Heat-treatable alloys can be solution heat treated by quenching immediately after brazing, and thus can be strengthened by aging alone.

Brazed aluminum assemblies have excellent corrosion resistance when properly cleaned of residual flux. Brazed aluminum joints generally resist corrosion as well as welded aluminum joints.

Brazed aluminum assemblies conduct heat and electricity uniformly. Brazed aluminum heat exchangers, evaporators, and similar complex fabrications are long lasting and highly efficient. The meniscus surface formed by the filler metal as it curves around corners is ideally shaped to resist fatigue.

Complex shapes with greatly varied sections are brazed with little distortion. Aluminum's excellent thermal conductivity assists in providing even distribution of the moderate temperature required for brazing.

Brazing makes precise joining comparatively simple. Unlike welding, in which the application of intense heat to small areas tends to move the parts out of alignment, parts joined by furnace and salt pot techniques are heated fairly evenly. Brazing makes part alignment easier. Brazed joints with tolerances of $\pm 0.002''$ are common in microwave component production.

Properly brazed joints are leak tight. For example, a vessel was sealed by brazing and evacuated to 2×10 torr for 100 hours. After that time, leakage increased internal pressure to only 1.6×10 torr, which is excellent for any metal joint.

Finishing costs are negligible. The capillary action that draws filler metal into a joint also forms smooth concave surfaces. Little mechanical finishing, if any, is required. When using a flux in brazing, removal of residual flux is required. The color match between parent metal and filler is generally good.

Personnel training is minimal. Production brazing equipment has been refined to the point where semiskilled and nonskilled people can perform most operations. Mechanically adept personnel can be trained in a few hours to torch braze.

PROCESS DESCRIPTION

The basic techniques employed to braze and solder aluminum are similar to those used to join other metals. In fact, the very same equipment used for brazing and soldering aluminum may be used to join other metals in this fashion, and this is frequently done so commercially.

Aluminum and its alloys have a number of physical properties that differ markedly from those of other metals commonly brazed and soldered. Aluminum's thermal conductivity is very high, it oxidizes rapidly, its coefficient of thermal expansion is greater than that of many other common metals, and aluminum does not change its color as its temperature changes.

The oxide that forms on aluminum as soon as the bare metal is exposed to air has a very high melting point — 3622°F. Aluminum oxide is neither melted nor reduced by temperatures that melt the metal itself. When aluminum is brazed or soldered, a flux is used to break up the oxide, float it away, and protect the bare metal base from further oxidation. Various mechanical means can be used to break up the oxide and expose the bare aluminum to the molten brazing filler metal.

The general procedure for atmospheric brazing aluminum is as follows. The surfaces to be joined are cleaned and spaced a few thousandths of an inch apart. A piece of brazing filler is positioned in or near the joint to be formed, and the joint is coated with a suitable flux. Heat is applied; the flux reacts, displaces the oxide on the surface of the base and filler metal, and shields the bare metal from contact with the air. The brazing metal filler melts and is drawn into the joint by capillary attraction. As the filler flows, it displaces the flux and wets the hot base metal, adapting to submicroscopic irregularities and dissolving the small high points it encounters. This takes place in a few seconds. After cooling and cleaning, the joint is ready for use.

Since the brazing fillers must melt at temperatures lower than the metals they join, their chemistries must be different. This results in a diffusion at the bond line which produces alloying in both the filler and base metal.

Aluminum alloys are brazed with filler metals similar to the base metals they are joining. The brazing filler metals have liquidus temperatures close to the solidus temperature of the parent metals. It is therefore very important to maintain close temperature control when brazing.

Liquidus and solidus are terms which define the melting zone of alloys. Whereas pure metals (and eutectics) melt and flow at the same temperature, alloys begin at one temperature (called the solidus) and are completely molten at a higher temperature (called the liquidus). Below the solidus temperature, the alloy is completely solid; above the liquidus temperature, the alloy is completely liquid; in between, the alloy is mushy.

ALUMINUM ALLOYS WHICH CAN BE BRAZED

Most of the nonheat-treatable aluminum alloys and many of the heat-treatable aluminum alloys can be brazed. The heat-treatable alloys most frequently brazed are 6061, 6063, and 6951. The nonheat-treatable alloys that respond best to brazing are 1100, 3004, 3003, and 5005.

Casting alloys which are brazable are A712.0, C712.0, D712.0, 356.0, 443.0, A356.0, 357.0, and 359.0.

ALUMINUM ALLOYS WHICH CANNOT BE BRAZED

Alloys 2011, 2014, 2017, 2024, and 7075 cannot be brazed with existing fillers. The melting points of these alloys are too low for the fillers developed thus far. Alloys with a magnesium content of 2% and greater are difficult to braze because present fluxes do not effectively remove the tenacious oxides that form on these alloys. High magnesium alloys of the 5000 series can be brazed by vacuum techniques.

BRAZING ALUMINUM TO OTHER METALS

Aluminum can be brazed to many other metals. A partial list includes the ferrous alloys and nickel, titanium, beryllium, kovar, monel, and inconel. Aluminum cannot be brazed directly to magnesium because an alloy forms which is so brittle it fails under very little stress.

Aluminum can be brazed to copper and brass, but the brittle compound which forms limits the joint's application. Aluminum can, however, be joined to copper and brass by means of a transition joint. Aluminum is first brazed to steel which is then brazed to the copper.

BRAZING METHODS

There are numerous brazing methods, but all include the same basic steps. The main differences between the various methods lie in the way the parts are heated and in the way flux and filler metal are applied. The only exceptions are vacuum and vibration brazing, which require no flux.

JOINT AND FIXTURE DESIGN

Simple, strong, and satisfactory brazed joints can be made by lapping one clean, flux-covered piece of aluminum with another. Brazing filler metal is positioned between or at the edge of the lap. The parts are heated until the filler metal melts and is drawn into the joint by capillary force. The joint is cooled, cleaned, inspected, and placed in service. Thousands of these simple joints are produced every day in the course of manufacture, field construction, and repair. However, when superior quality, tight tolerance, distortion-free, complex aluminum assemblies are to be brazed, careful designing is required.

BASIC DESIGN CONSIDERATIONS

Proper design normally begins with a careful study of the relationship between the brazed joints, the parts they are joining, and the dimensional criteria of the completed unit. During this study, the designer should bear the following parameters in mind.

1. The distance between faying surfaces (joint gap clearance) is important.
2. The coefficient of expansion of aluminum is almost twice that of the metals commonly used for the brazing fixture.
3. Aluminum is soft at brazing temperature and barely self-supporting. Thin sections tend to droop if unsupported.
4. Some distortion may be expected if a complex assembly is severely quenched after brazing.

Design anticipation can reduce the above considerations to negligible factors, for example, as follows.

1. Joint gap changes found necessary may be accommodated by using a lap joint. A change in joint clearance from the optimum generally results in lowered joint quality and should be avoided.
2. The difference between expansion rates of aluminum and a steel fixture amounts to about 0.005″ per 1″ at 1000°F. When small parts are brazed, this difference in expansion is often ignored. When larger, nonflexing parts

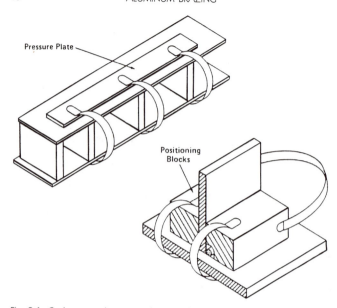

Fig. 2.1. C-clamp springs may be used to hold parts during brazing.

are brazed, springs, weights, and levers are used to hold parts in place. This aid also serves as an assembly fixture (see Fig. 2.1).

3. Small, thick, or vertical aluminum parts generally need no support during brazing. Long, thin, horizontal parts need support which may be supplied in many ways. Interlocking tabs, resistance welding, or tack welding may serve the purpose.

4. There are heat-treatable alloys which can be returned to their former temper without severe quenching. There are methods of quenching such as proprietary solutions, which reportedly preclude distortion. When distortion is expected and finish tolerances are closer than 0.005″, make the parts slightly oversize. Then the assembly may be finish machined after brazing. By this means, a brazed structure may be produced with close tolerances and flat sides.

JOINT PARAMETERS

The following conditions should be met if the highest quality brazed joints are desired.

1. Faying surfaces and a small distance beyond should be chemically clean and free of foreign adhesions and undesirable bumps and dimples. In other words, faying surfaces should be clean and flat. Generally, brazing shops merely degrease and braze.

2. The oxide, which is always present on aluminum, should be removed, just prior to brazing, from the faying surfaces and beyond for 0.5".

3. The distance between faying surfaces should be correct for joint width, filler and parent metals, brazing method, and time and temperature used.

4. The joint should be designed to allow entrance of flux and filler metal. Excess flux, oxide, and gases should be able to exit easily.

5. Faying surfaces and the adjacent fillet area should be fluxed prior to brazing unless the joint is to be dipped or vacuum brazed.

6. The assembly and furnace, if one is used, should be vented to allow escape of trapped expanded air and gases generated during the process.

7. The joint should be brought to proper temperature and held there long enough for a good brazing job to be done.

8. A fixed relationship should be maintained between parts during the process.

9. If a brazing sheet is supplying filler metal, the clad surface of the brazing sheet should touch the part it is to join for the length of the joint.

JOINT CLEARANCE

Capillary force is determined by joint clearance. Hundreds of simultaneous brazed joints are made possible by the drawing of molten filler metal into each crack and crevice, around corners, up vertical joints, and overhead. When capillary attraction acts on molten filler in a salt-dip brazing bath, it can draw liquid metal into and up a vertical joint 24″ high. This force makes it unnecessary to place filler in its final position. The molten flux reduces the filler's surface tension, allowing liquid filler to follow a clean fluxed path to its end, into a joint.

Suggested Joint Clearance for Various Brazing Methods

Dip Brazing and Vacuum Brazing	
Joint Width	Suggested Clearance
Less than 0.25″	0.002 to 0.004″
Over 0.25″	0.002 to 0.025″

Torch, Furnace, and Induction	
Less than 0.25″	0.004 to 0.008″
Over 0.25″	0.004 to 0.025″

Capillary force is directly related to joint clearance: the smaller the clearance, the greater the force. Also, the longer it takes molten metal to traverse the joint, the greater the possibility that oxide, flux, gas, and foreign material will be trapped inside. It is also possible that the filler will stop flowing prematurely since alloying with the base metal makes it less fluid at that temperature.

Overly large clearances pose problems of their own. Capillary action is reduced so flux and filler may not follow the joint to its end. More stress is placed on the fillet, and the joint will be weaker. Gaps may appear in the joint, a smooth fillet may not form, and filler metal will be wasted.

Fortunately, deciding on the correct clearance is easy. Satisfactory joint clearance may be determined by means of a few test joints made under actual production conditions. Once gap dimensions have been established, they will hold true as long as the other factors involved—brazing method, time, temperature, flux, and alloys—are not changed.

ESTABLISHING AND MAINTAINING JOINT GAP CLEARANCE

In casual brazing, joint clearance may be established by a layer of flux between the mating surfaces. A strong, useful joint will result if the assembly remains fairly motionless, except for small thermal movement during the cycle.

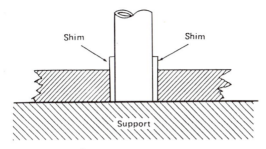

Fig. 2.2. Brazing filler metal shims both center parts and provide escape routes for gas and flux.

A simple but more positive method of establishing gap clearance is to use correct thickness shims of brazing filler to separate the parts (see Fig. 2.2). The shims melt during brazing, and they shrink about 5%. Upon cooling, they cannot be relied upon for close tolerance joint gaps. Additional support and fixturing are required.

IMPROVING BRAZING FILLER METAL FLOW

Molten metal flows best on clean, roughened or etched surfaces. Each scratch and pore acts as a capillary to pull the metal along. Grooved and serrated surfaces help to provide complete flow and a minimum of entrapped flux. Metal surfaces should not be smooth or mirror-like if they are to be brazed. There are proprietary chemicals for cleaning aluminum which also quickly etch aluminum surfaces that are too smooth.

AVOIDING FLUX ENTRAPMENT

Dry flux is chemically inert, but even in a dry joint there is no assurance that moisture will not enter and contaminate. This would, of course, initiate corrosion and damage. Good design provides an escape route for molten flux, heat-generated gases, and expanding air (see Fig. 2.3). It also allows easy access to solidified flux which must be removed after brazing.

Long joints should be vented along their lengths. Joints should not be filled with filler from both ends because that could lock flux and gas inside.

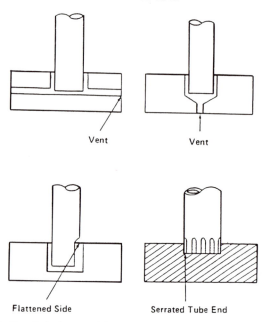

Fig. 2.3. Good design provides escape routes for molten flux, heat-generated gases, and expanding air.

Earlier, we mentioned that too small gaps could stop the flow of filler. They could also trap flux inside. Low brazing temperature or insufficient filler could also trap flux. At brazing temperature, flux releases hydrogen and other gases. At the same time, air within the enclosed space expands. Together they may produce pressures which would distort the assembly or even rupture it. Closed vessels must be vented. Afterwards, the vent may be brazed shut. Blind holes also must be vented.

JOINT TYPES

Refer to Fig. 2.4 for examples of brazed joint designs.

Lapped joints are strong, easily brazed, and require no special care. These joints can easily have more strength than the

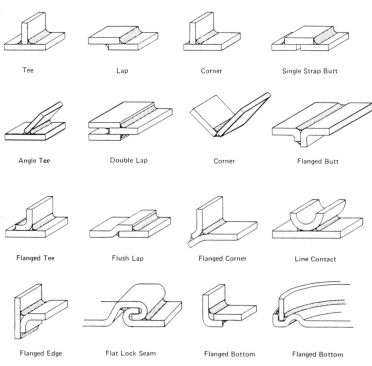

Tee	Lap	Corner	Single Strap Butt
Angle Tee	Double Lap	Corner	Flanged Butt
Flanged Tee	Flush Lap	Flanged Corner	Line Contact
Flanged Edge	Flat Lock Seam	Flanged Bottom	Flanged Bottom

Fig. 2.4. Brazed joint designs.

base metals by making the lap length two to three times the thickness of the thinner of the two base metals.

Lap joint clearance can be established and maintained in many ways. You could use a prick punch to dimple the thinner piece, or you could form protrusions of the correct height on the thicker part. Filler metal of the correct thickness could be positioned between the parts and a weight placed on top to hold the assembly immobile. Another method is to bend one of the pieces into an offset which may determine gap dimensions.

Tee joints are easily made since joint clearance is unnecessary if the vertical member is thin. The molten metal will generally flow under the edge of the vertical part bonding to the butting edge and forming fillets on both sides of the tee.

Line contact joints, such as occur when a tube or similar shape is positioned on a flat surface (as with a tee joint), are also easily made (see Fig. 2.5).

Strong butt joints can be made, but since they lack ductility, they should be avoided.

THICK METAL AND BROAD LAP JOINTS

When the vertical member of a tee joint is larger than ⅛″, the butting edge should be reduced somehow to improve filler flow. This can be done by rounding, serrating, roughing, or angling the butting edge. Any cut to this edge should not be larger than 0.01″.

Weep holes should be used to vent lap joints wider than ½″. When brazing aluminum, pressed joints should not be used

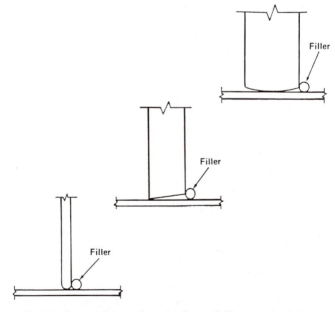

Fig. 2.5. Contact joints allow the flow of filler into the joint.

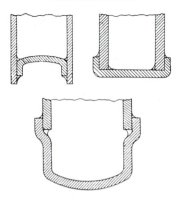

Fig. 2.6. Typical lap joints used for pressure vessels.

since there is danger of flux and gas entrapment. In fact, it is conceivable that the flux will not enter the joint at all.

Lap joints are best for pressure vessels (see Fig. 2.6) because they are the strongest type of joint; they provide the longest braze path, thus reducing leakage possibilities.

FILLER: PLACEMENT, SHAPE, AND QUANTITY

There are three ways to bring filler to the joint which is being filled. 1) Before brazing, the filler is placed in or near the joint. 2) Filler metal may be hand fed while brazing. 3) A brazing sheet can furnish filler. If the first method is selected, there are hundreds of rods, sheets, washers, tubes, wires, and rings. For hand feeding, filler is supplied in many wire sizes. For the third method, the sheet can be provided in many different thicknesses.

Generally, it is not necessary to measure filler metal exactly. Nevertheless, it is better to provide a little more than not enough. Using an excessive quantity of filler should be avoided because it is unsightly and could actually run where it is not wanted and interfere with operations.

Maximum fillet height can be achieved when dip brazing, that is, 0.5" high. With other brazing methods, the maximum fillet height is about 0.3".

If the assembly is removed too rapidly from a flux bath, or if it is shaken accidentally, fillet height can very likely be reduced.

PROPER FILLER PLACEMENT

A good design will have filler positioned as close to its ultimate resting place as possible. The shorter the travel distance for the filler, the less time there will be to alloy and the less oxide that has to be removed.

If the mass of filler metal is very small compared to the parts being brazed, the filler must be prevented from melting too soon. If it does, it may ball up and roll off. One step that may be taken to prevent this is to make filler grooves within a part.

Sometimes, when there is considerable mass difference between the parts being brazed, the thinner part will reach brazing temperature first. Therefore, it is best to place the filler in contact with the heavier part. It is imperative to protect filler metal from the source of direct heat when the heat is supplied from one direction; otherwise, the filler might melt prematurely and roll away before the base metal melts.

DIMENSIONAL CHANGES DURING THE BRAZING CYCLE

When brazing an aluminum plug in a hole in a piece of aluminum, if the part masses are similar, the hole and plug should expand and contract at about the rate that the temperature goes up and down. A correct joint clearance in this case, with cold parts, will provide a good assembly. If the external part is so large that it remains cool through the brazing cycle, the inner piece will expand and reduce the joint's clearance as it heats up. One inch of aluminum will expand about 0.012″ at brazing temperature (1000°). Steel, normally used for fixturing, expands about half that amount. This must be considered by the designer.

FIXTURING

Simple weights and levers will frequently be sufficient to hold parts in alignment except when dip brazing. Positive restraints are required when dip brazing. Although the media in a dip braze tank has a specific gravity similar to that of aluminum, and for all practical purposes supports the assembly, the parts are liable to separate as they are dipped into and removed from the bath. Self-fixturing assemblies are far less expensive to braze, although they may require more planning.

There are many costs which are caused by fixturing. First, there is the initial cost of design and fabrication. Next, there probably is a quantity of fixtures required for production. The fixtures pick up flux encrustations as do the assemblies and must be cleaned frequently. Fixtures add to both the heating and cooling load because they add to the assemblies mass. In dip brazing, fixtures contribute to contamination and add to the quantity of flux dragged out.

Parts can be held together by tabs, pins, crimps, stakes, rivets, springs, clips, tie wires, and welds. Only the designer's imagination limits the self-fixturing methods. If the brazed assembly sticks to the fixture, the parts could be coated with stop-off material, or the fixture may be constructed from oxidized stainless steel. Stainless steel fixtures can be oxidized simply by heating them, unloaded, in a furnace. Whatever fixtures are used, or whatever arrangement of weights, springs, or levers is devised, the aluminum pieces must be held lightly but firmly.

For short runs, low carbon steel is generally satisfactory for fixture material. If the flux is removed after each cycle, or at least often, the fixtures will last longer.

For production jobs, no bare or coated steel is recommended for dip brazing. Iron and steel will quickly contaminate the flux bath. The action of hot flux on nickel or aluminum coated steel will deteriorate these coatings in about 10 or 15 immersions.

It is advisable to use stainless steel or inconel X-750 for such assignments. For very long runs, inconel should be used because it will last much longer than stainless. With repeated exposure, the hot flux dissolves the nickel on the surface of stainless, and it is necessary to shot peen the dark color off the stainless (the dark color is exposed iron which would contaminate the bath). The fixture should be kept as small and light as possible to minimize heat loss.

APPLYING THE FLUX

All surfaces that are to be wetted by the filler, and the filler itself, must be fluxed unless, of course, one of the fluxless brazing methods or dip brazing is to be used. Molten filler will ball up unless fluxed.

Since flux is very hygroscopic, it should not be exposed to air longer than necessary; certainly not more than 45 minutes.

In other words, apply it as soon as possible before brazing. You want to assure that flux is present within the joint. Do this by covering the filler metal and both facing sides of the joint with flux. If the filler cannot be positioned in the joint, the path it will take must be fluxed.

Since steel containers would contaminate flux, it should be kept in special vessels (such as glass, porcelain, or porcelain-lined vessels). It may be applied by brushing or spraying. Flux in powder form may be mixed with water to form a thick paste. Fresh flux should be mixed every 4–6 hours.

Excessive flux should not be used since hot flux will stain any metal it touches. Yet sufficient flux must be used to allow it to do its job. When using a brazing sheet, the entire clad surface is normally fluxed to assist filler flow.

STOPPING FLOW OF FILLER

By eliminating flux from an area, the flow of filler can be stopped. However, preventing capillary attraction is a more positive method. This can be done by cutting short one of the brazing surfaces by 0.1″ where you want to stop the flow. There are many excellent commercial stop-off materials on the market (these are substances which block the flow of filler), and a few tests will indicate which stop-off is best for your job. The stop-off must not harm adjacent surfaces and it must wash off easily.

SEQUENTIAL JOINTS

When it is necessary to form a second joint adjacent to an existing one, a filler with a higher melting point must be used for the first joint. There are products made with a dozen adjacent joints. Each successive joint must be made with a filler having a melting point at least 50° lower than its predecessor. This is a common occurrence in metals other than aluminum.

HEAT-TREATABLE AND NONHEAT-TREATABLE ALLOYS

The physical and chemical properties of aluminum are altered by the addition of various elements. Certain elements produce alloys which can be hardened and strengthened only by cold work—these are known as nonheat-treatable alloys. The remainder of alloys, which can be strengthened by heat treatment, are called heat treatable.

Both groups lose strength as they are heated. The nonheat-treatable alloys soften and return immediately to O temper. The heat-treatable alloys must be maintained at annealing temperature for at least 20 minutes before they start to lose temper. These can be reheat treated. The nonheat-treatable alloys can only be tempered by more cold working.

Aluminum brazing temperature is between 1030°F and 1195°F, and annealing temperature is 650°F to 800°F. Brazing causes a slight annealing, and the amount of annealing which occurs depends upon the alloy, time/temperature, and mass. Recold working cannot often be done. Therefore, brazing a nonheat-treatable aluminum generally results in an assembly which is close to O temper.

In cases where high strength is not required, the selection of nonheat-treatable aluminum would reduce material cost. Most nonheat-treatable alloys (except those with high magnesium content) do not retain oxides as tenaciously as the heat treatable. When maximum quality is not required, nonheat-treatable alloys are often prepared for brazing by vapor degreasing alone—the chemical or mechanical cleaning is skipped.

When the aluminum leaves the quench tank, it is soft and workable. It can be kept in this workable condition for many hours by maintaining it at below freezing temperature. This can be helpful when there is much straightening to do and not enough time.

With the passage of time, these materials harden somewhat. If hardening occurs at room temperature, the alloy is said to have aged naturally; if hardening has been accelerated by heating, it is said to have been artificially aged. Time and temperature affect aging rate and temper, and vary from alloy to alloy. Some alloys take a month to fully self-harden, others require many months. Much can be learned about the heat treatment of aluminum by reading brochures from the aluminum prime producers.

An alloy which has already been hardened loses very little of its temper due to brazing. When some distortion is permissible, hot brazed parts can be dumped into boiling water to remove flux from the brazed areas. This is the best water temperature for flux removal.

Unless casting quality is very poor, castings may be brazed just as easily as wrought material. Brazable aluminum casting alloys include 356, 357, 359, 443, A712, C712, and D712.

DIP BRAZING

Dip furnaces are heated by AC current flowing between electrodes immersed in flux. Electrodes wear down and have to be replaced occasionally. Proper clearances must be maintained between electrodes and parts and between parts and the dip pot walls. Work which is self-jigging is ideal for dip brazing.

When one of the parts being brazed is clad (with solder), this also is an ideal candidate for dip brazing. One serious problem with this process is the fact that the dip braze pot must be kept fluid constantly. Consequently, you must have sufficient work to keep the pot working as continuously as possible, otherwise maintenance becomes too expensive.

FURNACE BRAZING

Except for dip brazing, more assemblies are brazed in a furnace than by any other means. It is so popular because the equipment cost is low and many existing furnaces can be used for brazing aluminum, then changed back again to another use. The weight of many assemblies can make them self-jigging for furnace brazing.

Furnace brazing is excellent for assemblies which might trap flux if dipped, or have pockets which could trap air. It is also good for highly polished parts which might get etched if dipped in hot flux.

Before putting parts in the furnace, they should be cleaned and have excess oxide removed. Faying surfaces are fluxed, filler metal is positioned, and the pieces are assembled and jigged. The assembly is then heated to about 300°F to drive moisture out of the flux. This is normally done in the preheat section of the furnace, perhaps on a conveyor. From there the assembly travels through the main portion of the furnace for about 3–5 minutes. The assembly is finally removed, cooled or quenched, then cleaned. Furnace time seldom exceeds 15 minutes.

TWO FURNACE SYSTEMS

Two types of furnaces are in common use for brazing today: the batch furnace and the continuous furnace. The cost of the batch furnace is low and its maintenance is also low cost. It can be run intermittently. Brazed assemblies can be made economically in batch furnaces. Cycling can be accomplished in 10–20 minutes, depending on the type of furnace and the mass of the parts. Batch furnaces are ideal for a large variety of parts per day and any production quantity whatsoever. Most any part which can fit in a furnace can be brazed.

The continuous furnace, on the other hand, could be one long unit or several joined units which normally have different temperatures in each section. The parts or assemblies move through the separate units on a conveyor. The parts are gradually brought to brazing temperature and finally moved through a cooling zone. This type of furnace can handle the highest hourly production rates and the highest hourly weights of aluminum of all methods of brazing.

When furnace time approaches 30 minutes, the quality of the brazed joints decreases. There are three basic reasons for this (in the order of their significance): flux change, liquation, and diffusion.

Flux change: When flux is exposed to air, it absorbs water and reacts with adjacent metal; beware of this hygroscopic characteristic of flux.

Liquation: When the temperature is raised very slowly, it could melt some of the lower melting point constituents of the filler material. That could change the chemistry of the filler and make it inadequate to do its job.

Diffusion: When assembly parts are very unequal in mass, the filler may turn liquid in one portion of the assembly before the rest of the assembly is ready for brazing. Nevertheless, any new job should be tried out because some assemblies which require much longer than 30 minutes can be brazed successfully.

SELECTION OF THE PROPER FURNACE

If your furnace can be brought to brazing temperature and maintained there within 5°F, it can be used for aluminum brazing. The heating source may be oil, gas, electrical resistance, or infrared heating lamps. Although any of these energy sources may be used for brazing, it is best to avoid the oil fired. Oil combustion is never complete, and soot may deposit on the aluminum parts. As a matter of fact, direct combustion furnaces, which are the least expensive, should be avoided. Moisture, always a byproduct of these furnaces, is likely to degrade brazed joints.

Hot-wall furnaces, in which hot combustion gases pass through tubes or behind furnace walls, provide a cleaner, dryer atmosphere for brazing. Alloys 6061, 6063, and 6951 do not braze well in direct combustion furnaces, so electrically heated hot-wall-type furnaces with unexposed heating elements are recommended.

TORCH BRAZING

The method generally used for one-of-a-kind brazing jobs, short production runs, and repairs is hand-held torch brazing. If the occasion requires it, the torch may be used for any brazing situation.

There are many sources of heat for torch brazing. However, most people use the same type of equipment, torch, controls, and gases which are used for fusion welding. For aluminum brazing, the operator merely changes goggle lenses and torch nozzles.

The art of torch brazing can be learned quickly and easily by most mechanically able people. Of all the gases used to fuel the torches, oxyacetylene produces the highest temperature. Other gases are cooler and thus easier to use in cases of light gage aluminum. By maintaining a low pressure, under 4 psi, the flame can be easily controlled. Visual inspection can be used to adjust oxyacetylene. The white cone produced by the mixture should be extended from the torch tip for 1" or 2", and the inner cone should be half this length. Oxygen pressure should always be about half of the accompanying gas. You should try to produce a flame which is slightly reducing. This type of flame tends to protect the aluminum parts from oxygen, since this flame consumes oxygen faster than the oxygen is delivered. A reducing flame is also less intense.

PROCEDURE

After the parts are cleaned, joint clearances are established. For torch brazing, joint gaps are normally set between 0.004″ and 0.025″. When possible, parts should be designed for self-jigging. At brazing temperature, aluminum is soft; the gas pressure exiting the torch could move it, so long sections need support. After flux is applied, care must be taken with application of the torch to prevent premature melting of filler.

MEASURING JOINT TEMPERATURE

There are three common methods of determining approximate joint temperature. 1) Flux indicates temperature by its appearance. As temperature increases, it dries out and turns white; with a continuing rising temperature, flux melts and turns gray. At brazing temperature, the flux becomes transparent and the aluminum glows with a silvery sheen. There are some proprietary fluxes which indicate temperature by showing color. 2) There are also special crayons which are used to make a mark near the joint. When the desired temperature is reached, the crayon mark changes color or melts. 3) The brazing filler itself is the third method. The tip of the filler wire is tentatively applied to the work being brazed from time to time. When the filler's tip softens and starts to melt, the brazing temperature has been reached. Temperature overshoot can be avoided at this time by withdrawing the flame for a moment.

HANDLING THE TORCH

The torch should be held a few inches from the work at any convenient angle from 5 to 45° to the part. Never permit the flame to remain on any one portion of the part for more than a moment; move it constantly either in a circular motion or from side to side. The outer edge of the cone provides maximum heat.

When parts of varying mass are being joined, the torch flame is moved rapidly over thin sections and more slowly over thick. Aluminum is such a good conductor of heat that heating unequal pieces is not as difficult as it may appear. When parts are very thin, it might be necessary to play most of the heat over the filler rather than the parts.

Supplying filler from one point is a good habit; this way, the filler will flow its own way down the joint. If the joint is too long, the filler may have to be fed from more than one position. Making sure that the bright metal flows the full length of the joint is a good way for the operator to ensure complete joints. If the joint is clean and fluxed, capillary action will draw molten filler into all cracks and form the fillet by itself. It is best to heat a joint once and not do any reheating because reheating of a fillet sometimes causes collapse of the matrix metal due to overheating.

When the heat of a single torch is insufficient to braze a large part, a second torch wielded by a second operator may be required. Multiple head torches are available to heat large assemblies. Assembly heat losses through conduction can be reduced by reducing contact between parts and fixture. Losses from convection or radiation may be reduced by reflective walls of some sort. OSHA frowns on the use of asbestos for that purpose.

When many assemblies have to be made, they can be brought to the operator on a conveyor. Fans are commonly used to cool brazed assemblies so that they can be quickly dropped into a wash tub for defluxing.

Since hot flux creates toxic fumes, brazing and cleaning should be done in a well-ventilated area. The operator should always wear protective clothing, gloves, and goggles. Try to wear the class and type of goggles recommended by the American Welding Society.

AUTOMATIC TORCH BRAZING

This is quite similar to manual torch brazing, except that the assembly or the torch is automatically moved, i.e., the torch may be programmed to move or the assembly may be moved past or around the torch.

First, the operator changes adjustments on the controls until the correct parameters for that specific braze are determined. When the torch-to-work distance, gas mixture, and heating time are correct, there is no need for further experiments. The flux and filler must also be positioned. Additional torches may be used to handle large masses.

In automatic brazing, care must be taken not to shake the assembly or rotate it too fast because of the danger of losing some of the fillets to gravity. Assemblies which cannot be dip brazed or furnace brazed often can be economically brazed by automatic torch.

DIP BRAZING

As in all other types of brazing, assemblies to be brazed by this method should be cleaned, freed of oxide, assembled, and jigged. Next, the assembly is preheated to almost 1000°F in a preheat furnace. Finally, the assembly is submerged in a dip-braze tub for one or two minutes. During this time period, one or many joints can be made simultaneously.

The hot molten flux is maintained at a temperature slightly above the liquidus point of the filler metal. It fluxes the joints, serves as a heat transfer agent, and partially supports the assemblies. Hot flux has the consistency of thickened water. It readily contacts every inch of surface and enters each joint and cranny unless prevented by dirt or air. The fluxing action is similar to the action of flux placed on a faying surface.

The temperature of dip-brazing furnaces can be closely maintained so that small, large, or heavy pieces can be consistently dipped without pause. Blind or inaccessible joints can be dip brazed easily and simultaneously.

THE SALT POT

Salt pots are steel-reinforced vessels lined with acid-proof fire brick about one foot thick. The large pot covers may be power operated on rollers, and they are also well insulated. These units generally last 10–20 years. To start the pot in operation, cold flux is placed in the pot and heated by a torch or portable electric heater. Once molten, the flux becomes electrically conductive and is kept at brazing temperature by the passage of electricity between electrodes immersed in the flux.

3

HARDCOATING

PROCESS OVERVIEW

Hardcoating gives an aluminum surface the hardness of case-hardened steel (Rc 50–60) while maintaining the light weight of aluminum. Hardcoated surfaces may be found in hospitals, schools, factories, in the home, under the sea, and on the moon.

When a permanent dry lubrication is added to the hardcoated surface, we find many more uses for it. A hardcoated aluminum prosthetic arm can have its joints permanently lubricated by this process.

Two types of hardcoats are currently available: the older type is formed at low temperature (32° F) in sulphuric acid; the newer type is formed at room temperature (60–70°F), also in sulphuric acid.

SEALING AND COLORING

Hardcoating is quite similar to anodizing but is about ten times as thick and much harder. Both anodizing and hardcoating processes form oxides on the surface of the treated parts. Very often these parts have the oxide covering sealed to prevent the absorption of undesirable liquids such as oil or colors or even perspiration from fingers.

Sometimes, when a hardcoated part is submerged in a hot solution (water or chemical) to seal the oxide covering the part, the large difference in heat conductivity between the aluminum matrix and the oxide will cause crazing. That is one reason the room temperature hardcoating method was created.

The conventional method of sealing either anodized or hardcoated aluminum is immersion in distilled or deionized water at or near boiling. In the case of dyed films, a bath of nickel acetate, at the same temperature range, is used. In addition to the possibility of causing crazing, sealing can also degrade the abrasion resistance of hardcoating.

Coloring hardcoated surfaces is similar to coloring anodized surfaces. It is done with the same dyes and procedures. The main difference is in the depth of colors. The oxide of anodized parts is much more porous so it absorbs more color.

Recently, products for sealing have come on the market that work at or near room temperature. These products should prevent crazing problems. The subject of sealing and coloring should be discussed and evaluated with the processing company whenever hardcoating is considered because vendors, who have been on the scene for 10–20 years, can give the best advice.

Note that hardcoated surfaces do not have to be sealed. In fact, there is a military specification covering this.

FINISHING

Sometimes it may be necessary to lightly grind a hardcoated surface to make it flat or parallel. There are many abrasive wheels and compounds suitable for finishing hardcoated surfaces to achieve critical dimensional tolerances or very fine finishes.

For surface grinding, silicon carbide grit, size 80–120, will provide a finish of 8–2 microinch. For cylindrical grinding, a finer grit wheel, Norton 39C120-J8VK, will be free cutting and yet produce a fine finish. For internal grinding, a fine grit wheel, such as Norton 39C100-J8VK, produces the best results. In general, grinding should be done wet using a water coolant and a good soluble oil mixed approximately 100 to 1. For polishing or lapping, a boron carbide abrasive grain mixed with oil will give good results. The range of grit size should be 400–1200 depending on the finish required.

WROUGHT ALLOYS

Hardcoating is recommended for use with virtually all aluminum alloys. It is important to remember that hardcoat thicknesses are 50% penetration and 50% buildup.

1100 Series—The 1100 is very common. Bronze-gray in color at 0.002″ thickness. The alloy is difficult to machine and is soft.

2000 Series—The 2014, 2017, 2024, and 2618 forgings are very common. Avoid sharp corners. Gray-black in color at 0.002″ thickness to blue-gray at 0.004″ Good machinability. Thickness to 0.006″ for salvage, although not as hard as thinner coats.

3000 Series—Most common is 3003. Gray-black in color at 0.002″ thickness. Machines easily and is good for dye work.

4000 Series—Not commonly used.

5000 Series—Most common are 5005 and 5052. The 5005 is better for dye work; 5052 accepts only black. Both machine well. The 5052 gets excellent dielectric value when coated 0.004″ thick.

6000 Series—The 6061 and 6063 are most common. Almost black at 0.002″ thickness. The 6061 forms excellent hardcoat for grinding or lapping. Good dimensional stability. The 6063 is used for extrusions.

7000 Series—Most common is 7075. Blue-gray at 0.002″ thickness. High strength alloy. Not good for grinding or lapping. Maximum for salvage is 0.008″ thick.

INGOTS

Sandcast Alloys—Most common are 319, 355, and 356 (also 40E, Ternalloy, Tenzalloy, and many other proprietary alloys). The 356-T6 is the most popular. Good for grinding and lapping. Hardcoat will not fill in exposed surface porosity which is common with sandcastings. Vacuum impregnation (plastic) will improve the hardcoated finish of a sandcasting.

Diecast Alloys—Most common are 218.360 and 380. Only 218 produces hardcoat comparable to that on wrought or sandcast material. However, 218 is difficult to diecast. Maximum thickness is 0.0025″. Maximum thickness for 360 and 380 is about 0.001″.

The reason for the difference in hardcoating quality is the alloys. Silicon and copper are detrimental to a good hardcoat.

HOW TO ORDER HARDCOATING

To save time and trouble when ordering hardcoating, information on the following four items must be known: 1) alloy, 2) coating thickness, 3) masking requirements (if any), and 4) racking instructions (if possible).

1. Alloy—Since the hardcoat builds up at different rates on each alloy, it is important to specify the alloy. Also, some alloys require procedures which vary slightly from one another.

2. Coating Thickness—Hardcoating can be provided in a range of thicknesses from 0.0002″ to 0.009″, depending on the alloy and the application. Since half the coating is penetration, and only half adds to the thickness, this change in dimension must be taken into account on the blueprint.

3. Masking—If it is necessary to exclude (mask) certain areas from hardcoating, they should be clearly indicated on the blueprint. Generally, the need to mask adds cost to the job. If it is possible to hardcoat all over, this is the way to proceed unless this would require using oversize taps.

4. Racking—Each part being hardcoated must have a good mechanical and electrical contact, called racking. Each racking point leaves a small void in the coating, and these voids should be made in a noncritical area.

FACTS ABOUT HARDCOATING

Hardcoating is different from plating. For example, hardcoating a shaft 0.002″ thick will increase the diameter by only 0.002″, whereas plating the shaft 0.002″ thick will increase the diameter 0.004″.

When hardcoating is called out on a drawing, the added phrase "build up per surface" should be included to prevent a misunderstanding.

Standard commercial tolerance is 0.0005″ on a coating thickness of 0.002″.

To allow for hardcoating a "V" thread, the formula is "Build up per surface multiplied by four."

Hardcoating is not compatible with anodizing. Parts may be damaged if anodized after hardcoating. When there is a requirement for hardcoat and any other type of chemical processing such as anodize, alodine, or irridite, contact the hardcoat vendor for advice.

Remember to specify teflon impregnation with the hardcoat if the highest lubricity is desired.

4

TITANIUM NITRIDE COATING FOR SUPERIOR TOOL PERFORMANCE

GENERAL INFORMATION

In these days of intense industrial competition, the cost savings and productivity improvements associated with titanium nitride coating of cutting tools are too important to ignore. If tool engineers want to act aggressively, they would use this coating in every situation possible.

A titanium nitride (TiN) coating has a hardness of about 80 Rc and a coefficient of friction about one-third that of the steel substrate. The coating protects tools from abrasive wear, and the lubricity reduces friction, heat, and the buildup which forms on tool edges during use.

TiN-coated tools generally pay for themselves the first time they are used. The coating reduces cutting forces by about 17% in most drilling applications, and increases tool life 4–9 times in cast iron, stainless steel, nonferrous metals, and carbon steels. Even after a TiN-coated cutting tool is resharpened, the tool life is still twice that of an uncoated tool. Coated tools are sharpened the same way as uncoated. TiN-coated taps reduce cold welding when tapping mild steel and reduce friction when tapping stainless.

TiN coating on thread rolling and similar tools, which move metal rather than cut it, have the same saving potential. That is also true for countersinks, counterbores, reamers, end mills, milling cutters, hobs, broaches, turning tools, and saws. It may be necessary, on some occasions, to change grinding angles. When the depth of cut is slight, it may occasionally require experimentation. The cutting edge may not bite into the work the same way an uncoated tool would. Coating vendors can advise when adjustments are required.

When punching galvanized steel, carbon steels, bronze, and soft iron, longer tool life is reported when the punches are coated. The punches have less material buildup on their sides and less galling. Stripping action is also improved.

In semiwarm operations, excellent results can be expected. But coating forging dies is not recommended. Punches and forming tools made of D2 tool steel should be heat treated with 3 tempers at 950° F; this ensures dimensional stability and no loss of hardness when the tool is subjected to the coating process temperature.

Applying TiN to molds is very beneficial. Material buildup, like rubber or plastic, is easier to remove.

The best tool surface for coating is finely ground, clean, and with no prior treatment. That means no nitride, black oxide, or any other surface treatment. A coating vendor will have to advise if you have doubts. Before polishing molds or any other tools, check with the coating vendor because some polishing materials are difficult to remove, and the vendor will probably recommend some water-based polishing solutions.

PROCESS DESCRIPTION

In this coating process, the tools to be coated must be physically and chemically clean. They are positioned in fixtures which are the cathode of a high voltage system in a reaction chamber. The chamber is evacuated and charged with argon for a process called sputter cleaning. The positive argon ions are propelled by a high voltage field to blast the tools so they become atomically clean. Titanium is heated by electron beam until it evaporates. At this time, nitrogen is introduced into the chamber and the titanium ions are electrically accelerated toward the tools. The titanium ion bombardment combines with the nitrogen gas to form a coating of titanium nitride about 0.0001″ thick on the exposed surfaces of the tools. The adhesive bond between the coating and the matrix is so tight, the coating does not separate due to deflection when the tools are used.

The coating process described above, called "Physical Vapor Deposition" (PVD), operates at about 900°, which is well below the tempering temperature of high speed steel. That means hardened tools will not soften during the coating process. There is another coating process, called "Chemical Vapor Deposition" (CVD), which operates at a temperature of 1750–1950°F. It was originally developed to coat carbide inserts; however, if tool steel is coated by this process, it would soften and require rehardening. For this reason, the PVD process is normally used for coating tools. The performance of carbide tools coated by PVD is improved because the process minimizes edge embrittlement. The disadvantage of CVD is that it tends to round over sharp edges and thicken corners.

BREAK-IN PERIOD

When cutting tools are first used, they develop a wear land during the break-in period. The wear of the cutting edge then proceeds at a much slower pace; this is called the low wear period. The useful life of the tool continues until rapid edge breakdown begins to occur. At this time, the tool should be sharpened. When a tool is titanium nitride coated, the low wear period is extended far beyond that of uncoated tools. Forming tools, plastic molds, and mold components coated with titanium nitride show similar reduced wear characteristics.

There are numerous reports detailing dramatic improvements in tool life and productivity due to the use of TiN-coated drills, end mills, milling cutters, and other high speed steel tools. TiN-coated tools are now used on all metal removal operations to extend tool life, increase feeds and speeds, and reduce machine downtime.

Three areas should be considered by those who would like to understand the advantages of coated tools: the tool quality, the consistency of the quality, and the tool performance.

TOOL PREPARATION

Any rational evaluation of TiN-coated tools must begin with consideration of the tool prior to coating, i.e., it must be of good quality to achieve superior results after coating. The condition of the tool's cutting edge is a prime example. A ragged cutting edge and heavy burrs on any tool would waste the cost of coating. Coated tools with poor quality surface finishes would probably not show worthwhile improvement after coating.

Rough grinding will leave edges that break off easily when cutting starts. That would leave uncoated surfaces exposed, to wear rapidly. Tools with grinding burns, rust and surface oxidation, or previous surface treatment would have inferior adhesion of the TiN to the tool. Premium high speed steel, when coated, will show better results than general purpose HSS.

It is essential that all procedures in the coating process be carefully monitored if consistent quality is to be maintained. All tools should be evaluated at incoming to establish their suitability for coating. Each tool should also be evaluated for the correct cleaning procedure at this time. Chemical, electrical, or mechanical steps are possible.

All the operating procedures—from arrival of the tool at incoming, to packaging the tool for return—are significant. All the coating and inspection machinery should be maintained in good condition. The final quality of the coating job is liable to be affected by any step in the procedure—even housekeeping. This type of work shows better results when "clean room" conditions are maintained.

Managers should do everything necessary to ensure that the coated tool is used in a manner which will derive the highest return. This means that proper machining practices should be followed. The workpiece should be fixtured rigidly; tool holders, arbors, and any other aids used should be in good condition for running tools at high speeds and feeds. The full value of the coated tool will not be realized unless feeds and speeds are gradually increased to their maximum efficiency.

The tool should be resharpened before the coating is seriously damaged. This step will allow less stock removal and increase the number of times resharpening can occur. At some point, the tool should be recoated.

MATERIALS GUIDE FOR TiN PROCESSING

The following materials can be coated.

All heat-treated HSS grades, including cobalt materials and CPM powdered metals.

Heat-treated tool and die steels.

Solid carbides.

Carbide tipped tools (providing the brazing fillers are free of zinc or cadmium).

Most grades of stainless steel.

The following materials cannot be titanium nitride coated.

Assemblies which are bonded, pinned, or screwed together permanently.

Certain aluminum alloys.

Any alloy containing lead, zinc, tin, or cadmium.

Surfaces best for titanium nitride coating are as follows.

Finely ground surfaces, for maximum coating adhesion.

Surfaces free of burns or grinding wheel glazing.

Surfaces resulting from the use of free cutting grinding wheels to grind with lower temperatures.

Tool surfaces that can be coated after special treatment, as follows.

Milled or machined surfaces.

Nitrided surfaces.

Polished or lapped surfaces.

Black oxide surfaces.

EDM machined surfaces.

The following surfaces require special treatment before coating.

Painted or marked surfaces and any covered with plastic.

Chrome plated surfaces.

Rusty surfaces.

5

POLY-OND

A NEW TECHNOLOGY FOR PLATING METAL

Poly-Ond, a proprietary formulation developed in 1976, is a liquid bath process which chemically deposits nickel phosphorus, impregnated with polymers, on the surface of metal parts. This process permits the use of less expensive metals when anticorrosive materials are called for.

ENGINEERING SPECIFICATIONS

Hardness as deposited—Rc50.

Hardness after a 1 hour bake at 750°F—Rc68–70.

Coefficient of friction under 200 pound kinetic load—0.06.

Salt spray resistance per ASTM-B-117 test—300 hours.

Thickness range per coating—0.0002–0.003" per surface.

Standard plating thickness—0.0005" + or − 0.0001".

Standard bake temperature—700–750°F.

Thickness controllability— + or − 0.0001".

Operating temperature range—freezing to 550°F.

Poly-Ond plating has the approval of the United States Department of Agriculture for applications on food processing equipment.

Size for processing is from watch parts to 5 tons in weight, and up to 30" diameter and 12 feet in length.

LUBRICITY

The process creates an infusion of polymers throughout the thickness of the coating. As the coating wears, there is a continuation of lubricity. The lubricity achieves an exceptional release for all types of molds, and is commonly used in applications where a low coefficient of friction is required in addition to hardness of material.

COST INFORMATION

At present, the approximate cost of this process can vary from 35–50¢/square inch depending on the requirements of stripping and/or masking. Of course, there is a minimum charge of about $50 for any order. Any type of plating would ordinarily have a similar charge. In fact, to be thorough when considering the cost question, a comparison should be made to the cost of competitive finishes.

ADDITIONAL TECHNICAL INFORMATION

Avoid masking if possible. It is often less expensive to coat an entire surface rather than mask areas.

If the part material will not allow the standard processing temperature, discuss the situation with the vendor because processing gains are possible at 250°F.

When two Poly-Onded surfaces run against each other, it is best if their hardness is mismatched by 10 Rc points. This type of mismatch should be used whatever finish is on your parts.

Most standard manufacturing materials can be coated, including tool steels, stainless steels, aluminum, copper, and bronze.

This is an electroless process of chemical reduction, and as such will deposit very evenly.

USES OF THE PROCESS

The designer should be aware of the different types of surface hardening available. There have been cases when one type did not work well for very long and another was substituted quite successfully. The material (matrix) is often critical in making a process selection; and sometimes the function is critical.

Recently, a heavy assembly fixture was being moved back and forth across a granite surface plate. It moved on three standard fixture feet, each ¾" in diameter. After a short period of time, there were marks scratched by the feet in the granite. This, of course, required expensive repair work. Before the fixture was used again, the three feet were Poly-Onded, and a repetition of the same wear problem was avoided.

Notes should be kept by design personnel of successes and failures in their choices of surface finishes because there is no substitute for actual experience.

ELECTROLIZING

DESCRIPTION

The electrolizing process uniformly deposits a very dense, high chromium, nonmagnetic, hard (70–72 Rc), proprietary alloy onto the part being treated. This alloy provides an unusual combination of properties: wear resistance, low coefficient of friction, good antiseizure characteristics, excellent corrosion protection, and a good sealant to the covered surfaces.

Electrolizing involves cleaning the part by removing a surface layer from the base metal, and following with an electroplate which positively bonds the chromium alloy to the porosity of the base metal.

ADVANTAGES

Some desirable features of this process are as follows.

Precision thickness from 0.000050" to 0.001".

Absolute adhesion of plating to matrix.

Superior corrosion protection.

Coating is conductive: eliminates electrostatic buildup.

No outgassing.

Stain resistance: the color will be satin-gray, smooth, fine-grained. Most matrices will finish with the same color.

Coating can be applied to both heat-treated and nonheat-treated tool steels, all stainless steels, all nickel alloy materials, all aluminum alloys, brass, copper, bronze, and titanium, and all commonly machined ferrous and nonferrous metals.

GENERAL INFORMATION

Electrolizing should be done after all metal processing is finished; that includes stress relieving or straightening.

The coating should be applied directly to aluminum surfaces without intermediate coatings of copper or nickel.

Electrolizing is not intended as a substitute for heat treatment.

The adhesion should be such that when viewed at 4 diameters, it will not show separation from the base metal on test specimens bent through 180 degrees on a diameter equal to the thickness of the specimen.

Electrolizing will generally not be affected by thermal shock. In fact, it will adhere at temperatures ranging from $-500°F$ up to $1600°F$.

Size limitations are 10 feet in length, 20 inches in diameter, and 1000 pounds in weight.

The coating is not recommended for beryllium, magnesium, columbium, lead, and their respective alloys.

MOLDING PROBLEMS

The plastic and rubber molding industries are plagued with problems which are related to platings or coatings. There is a growing use of tougher resins filled with glass, minerals, flame retardants, and other abrasive and corrosive additives. This means that processing equipment components like barrels, screws, nozzles, mold cavities, pins, runners, and mold plates will wear rapidly unless they are protected with a coating process like electrolizing (see Fig. 6.1). Note that areas may be masked if required.

ATTRIBUTES OF THE PROCESS

Electrolizing can prevent or retard a number of harmful occurrences. For example, galling, caused by a pile-up or gouge-out of metal, can lead to seizure, which is the welding of one metal to another. Scoring occurs when small particles scratch the metal during movement.

Fig. 6.1. Injection mold parts protected with electrolizing coating.
(Courtesy of Electrolizing Inc.)

Electrolizing also protects against corrosion. Surface corrosion of metal is a common problem which always must be considered in manufacturing. It is caused when unprotected metal
is exposed to chemicals, water, or even moisture in the atmosphere. If metal surfaces are left unprotected, surface erosion and
pitting of the surface will result; then the integrity of the metal
must be restored by remachining or repolishing.

Friction and heat caused by metals rubbing together cause
wear. Electrolizing with its high surface hardness (Rc 72) and
low coefficient of friction, increases wear life (see Fig. 6.2).

Fig. 6.2. Electrolizing, with its high surface hardness and low coefficient of friction, is used to increase wear life of aircraft parts. (Courtesy of Electrolizing Inc.)

It is interesting to note that most geometries can be uniformly coated. However, when slots, grooves, and even threads are in your part, the vendor should be consulted.

The processing temperature will not exceed 180°F which precludes heat damage and hydrogen embrittlement.

Aluminum alloys are generally anodized or hard coated because this treatment, called conversion coating, is preferred.

That is because aluminum has a high affinity for oxygen which makes electroplating difficult. Most metals used to plate aluminum are cathodic to aluminum, which means that any voids in the coating, no matter how small, will lead to localized galvanic corrosion. Also, the thermal expansions of aluminum and most metals used in electroplating are so different that it could result in flaking and nonadherence. Electrolizing, on the other hand, is an ideal coating process for aluminum, and this has been proven through countless applications over the years.

7

POWDER COATING

HISTORY

World War II created, in Europe, a shortage of solvents required for conventional painting operations. Dr. Irwin Gemmer, a German engineer, discovered a way around this shortage. He found that when heated metal parts were dipped into a cloud of powdered plastic suspended by air turbulence, the particles would melt and cling to the part. He then discovered that when these coated parts were heated and cooled, the plastic particles fused and solidified into a tough, durable, totally encapsulated finish with no drip marks. This was the birth of the powder coating industry.

After 10 years, it was learned that charging the plastic particles electrostatically would also make them cling to the workpieces. The electrostatic method, Gemmer's technique (now called fluidized bed), or a combination of the two are the main methods still being used today. The techniques have matured, and they now represent a superior finishing process.

USES

Electroplating, anodizing, bonderizing, phosphating, and painting have all lost business to powder coating. Powder coating does not affect the rigidity or strength of the matrix material. When a part is powder coated, it will resist weathering, abrasion, powerful chemicals, and electrical charge. The coating may also be used for decorative purposes.

These coatings protect personnel from sharp edges and rough surfaces, and, in many cases, reduce polishing, grinding, and deburring efforts. Wire fabrications, castings, forgings, stamp-

ings, springs, and subassemblies have all been powder coated successfully, saving considerable funds in the process. Size or shape or complexity are not limiting factors.

There are good reasons for utilizing powder coating: abrasion resistance, personnel protection, ultraviolet resistance, electrical insulation, thermal insulation, corrosion protection, sound deadening, chemical resistance, and pleasant feel to the touch.

MATERIALS

The range of plastic powder available for coatings is constantly growing, although there are many powders which are used only occasionally because of special characteristics. At present, there are five families of powder commonly used: nylon, vinyl, epoxy, polyethylene, and polyester.

NYLON

Nylon is the conventional choice when a low coefficient of friction or a tough plastic is desired. It is also excellent when abrasion or resistance to solvents and alkalis is a problem. Nylon is autoclavable.

VINYL

Vinyl performs very well when superior resilience and waterproofing are needed. The vinyl surface cleans easily, is nonslip, and is resistant to alcohols, oils, inks, and foods. Vinyls are tasteless and nontoxic. Although vinyls are soft and attractive, they have good impact resistance. Vinyl withstands salt spray and weathering.

EPOXY

Epoxies provide a hard coat with excellent temperature resistance. They possess unusual adhesion to most surfaces, excellent toughness, strength, and thermal shock resistance. They have a full range of colors and are resistant to salt spray.

POLYETHYLENE

Polyethylene exhibits chemical resistance, inertness, flexibility, and electrical insulation. It is soft and warm to the touch. It has good impact strength and low cost.

POLYESTER

Polyesters are hard, tough, and have excellent adhesion and impact strength, so they are probably the best choice for weather resistance. They have a full range of colors, textures, and glosses.

The following is a partial list of plastics in general use, but which are not used as commonly as the above-mentioned five:

Fluorocarbons

Polyurethanes

Polypropylenes

Acrylics

Cellulose acetate butyrate.

Another type of plastic coating, which is vital in numerous mechanical designs, is the organic coating. When your part requires some additional characteristic, such as lubricity, abrasion resistance, or chemical resistance, there are several coatings which may be used. We have had excellent results using six of the organic coatings listed in the Precision Coating Company's charts (See Table 7.1) and recommend their consideration. These plastics are generally sprayed over the parts and then cured as described.

COST

The cost of powder coating varies widely depending upon the plastic used, the surface area covered, and the thickness desired. A high production rate would allow automation to be used. If powder coating cost is investigated in a manner similar to that used when considering plating or painting, the price would be very competitive.

METHODS

Earlier, mention was made of the fluidized bed and the electrostatic methods of applying plastic coats. These are the two most common methods, and most coating vendors use them. However, there are companies which create their own methods. Plastonics Inc. of Hartford, CT, advertizes five powder coating methods, which are described and pictured below.

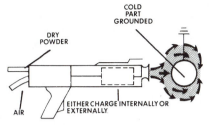

Fig. 7.1. Electrostatic coating. Cold parts are coated with unheated
electrostatically charged dry plastic powder.

ELECTROSTATIC COATING

In this process, cold parts are coated with unheated electrostatically charged plastic powders which are sprayed on. This
method is well suited to the application of powder to localized
areas as well as to complete parts, and thin coatings may
be applied. Many plastics in many colors are available. (See
Fig. 7.1.)

ELECTROSTATIC FLUID BED COATING

This process involves passing unheated parts over an electrostatically charged fluidized bed of dry plastic powder. The
method is excellent for small parts and where thin coats are
wanted. It is recommended where masking of areas is required.
(See Fig. 7.2.)

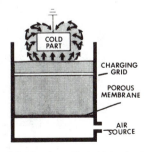

Fig. 7.2. Electrostatic fluid bed coating. Unheated parts are passed
over an electrostatically fluidized bed of dry plastic powder.

Table 7.1. Organic Coatings

	Coating System	Outstanding Properties	Continuous Operating Temperature	Normal Thickness Range[b]	Approximate Buildup Per Coat[a]	Curing Temperature	F.D.A. Acceptability
Phenolic-Epoxies	PC-11	Excellent dielectric Good abrasion resistance Good corrosion resistance	350°F	2.0–5.0 mil	1.0 mil	375°F	No
	PC-48T	Good corrosion resistance Good flexibility Flat black in color	350°F	0.5–1.5 mil	0.5 mil	350°F	No
	PC-7100	Good corrosion resistance Good abrasion resistance Good dielectric	325°F	2.0–5.0 mil	1.0 mil	ambient to 300°F	No
	Bisonite Phenolflex®	Excellent corrosion resistance Good abrasion resistance	350°F	2.0–5.0 mil	0.7 mil	400°F	Yes
	Plasite®	Excellent corrosion resistance Good abrasion resistance	300–400°F	2.0–8.0 mil	1.0 mil	ambient to 400°F	Yes
Dry Film Lubricants	Electrofilm®	Excellent lubricant[a] Molybdenum disulfide base	500–1000°F	0.2–0.5 mil	0.3 mil	400°F	No
	Everlube®	Excellent lubricant[a] Molybdenum disulfide base	500–1000°F	0.2–0.5 mil	0.3 mil	400°F	No
	Lifelube® LLC-30	Good lubricant[a] Molybdenum disulfide base	400–500°F	0.3–0.5 mil	0.3 mil	400°F	No
	Moly-Dag®	Excellent lubricant[a] Molybdenum disulfide and graphite base	275–350°F	0.3–0.7 mil	0.3 mil	300–400°F	No
	Molykote®	Good lubricant[a] Molybdenum disulfide base	400–600°F	0.3–0.5 mil	0.3 mil	ambient to 350°F	No
	Teflon®	Excellent lubricant[a] Teflon base	300–500°F	0.5–1.0 mil	0.5 mil	450–700°F	No

	Properties					
Teflon® T.F.E.	Excellent hi-temp resistance / Excellent release / Available in black for draining static charges	500°F	1.0–5.0 mil	0.5 mil	700°F	Yes
Ryton® T.F.E.	Excellent hi-temp resistance / Good abrasion resistance / Good release	450°F	0.7–2.0 mil	0.5 mil	700°F	Yes
P.C. 8-403	Best combination of release and wear	350°F	1.0–1.5 mil	0.5 mil	700°F	No
Teflon® S + F.E.P.	Good release / Good abrasion resistance / Good chemical resistance	350°F	1.0–2.5 mil	0.5 mil	700°F	Yes
Teflon® F.E.P.	Excellent chemical resistance / Good release	350°F	1.0–4.0 mil	0.5 mil	700°F	Yes
Kel-F®	Excellent abrasion resistance / Good chemical resistance	350°F	2.0–10.0 mil	0.7 mil	500°F	No
P.F.A.®	Good abrasion resistance / Excellent chemical resistance / Good release	500°F	1.0–5.0 mil	0.7 mil	700°F	No
Teflon S®	Excellent abrasion resistance / Low friction	325–400°F	0.5–1.5 mil	0.5 mil	450–600°F	Yes
Emralon®	Good abrasion resistance / Low friction	325–400°F	1.0–2.0 mil	0.5 mil	300–400°F	No
Xylan®	Widest color range / Low cure / Low friction	500°F	1.0–2.5 mil	0.7 mil	250–400°F	No

Fluorocarbons

Courtesy Precision Coating Co., Inc.   [a]Meets certain military specifications.   [b]mil = 0.001 in.

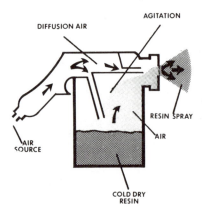

Fig. 7.3. Flocking.

FLOCKING

This method is employed when small quantities are needed. The process is similar to the electrostatic fluid bed, except that the parts must be preheated and no electrical charge is used. (See Fig. 7.3.)

FLUIDIZED BED COATING

This process involves the preheating of metal parts and their immersion into a fluidized bed tank filled with dry, plastic powder. Components or assemblies coated by this process can be handled in small or large quantities. The process is very economical. Heavy coats of a wide variety of materials are used. (See Fig. 7.4.)

FLO-CLAD

This method is generally limited to very small parts not exceeding three ounces in weight, three inches in length, and requiring processing of large quantities. (See Fig. 7.5.)

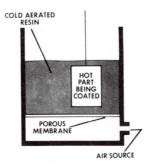

Fig. 7.4. Fluidized bed coating. Preheated parts are immersed into a fluidized tank filled with dry plastic powder.

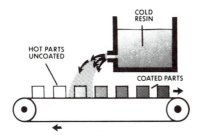

Fig. 7.5. Flo-clad.

GENERAL FUNCTION

A multitude of problems has been solved by plastic coating, and many functions have been improved by introducing the process. For example, unsightly welds can be hidden under plastic coating, and minor surface imperfections obliterated. A battery manufacturer found that plastics were much more resistant to acids than paint. Also, many pieces of hardware are improved functionally and aesthetically by coatings.

SURFACE COATING RECOMMENDATIONS

In the preceding chapters, the following surface finishing techniques were discussed:

Titanium Nitride Coating
Plastic Coating
Hardcoating
Electrolizing
Poly-Ond.

All these coatings have advantages for specific applications, and they all pass military specifications in their categories. If you contemplate using any of them, it is recommended that you discuss the situation with your vendors; then you can establish your own criteria and inspect the results accordingly. By comparing the costs of the process with their relative results, you can establish your own values for each.

For instance, titanium nitride coating stands alone in that it is the only process the author can recommend for coating cutting and forming tools. There are probably several vendors capable of applying some coating to do the same job well, including TiN by Balzer Inc., a supplier of high-quality coatings. (They have three locations, one of which is North Tonawanda, NY.)

Plastic coating has a unique place in industry because it is used more as a decorative finish than for lubricity or hardness. It also protects against corrosion or scratching and cutting. Its main competition comes from painting.

The third process, hardcoating, is only for aluminum. It is the traditional method for hard surfacing aluminum, and, when followed by teflon impregnation, it is indeed a very useful design tool.

Now, here is where the serious competition begins. Electrolizing and Poly-Ond are both excellent processes for finishing aluminum: both are hard coats (Rc 50–70) and both have good lubricity; they are both outstanding for finishing ferrous materials as well as aluminum; and both are proprietary, which means there is a limit to the information obtainable. Hardcoating, electrolizing, and Poly-Ond are all reliable; you will simply have to compare results with costs in order to make your choice.

8

ELECTROFORMING

PROCESS DESCRIPTION

This process has been available for decades, but it is still considered a space-age technology. In electroforming, a metal is electroplated onto a part, the part is removed and thrown away, and the plating remains. Actually, the techniques of electroforming are not quite so simple; they differ considerably from ordinary electroplating, yet the family resemblance remains. The basic capabilities and chemistry of plating, known collectively as Faraday's Law, govern electroforming.

Formerly, this process was practiced as an art and, accordingly, provided erratic results. What is new is the application of advanced technology to the field. The current increased emphasis on precise, thin-walled metal components has placed a new light on this process. Waveguides, optical quality reflective surfaces, and even aerospace components are now being produced on a production basis. Electroformed dies are lowering the cost of some intricate dies for both metal and plastics. Now, a better understanding of electrochemistry, and the role of chemical additives in plating baths, permits much closer control of electroformed parts. Results are now as reproducible as in casting, forging, and other conventional processes.

As in electroplating, metal ions are transferred through an electrolyte, from an anode to a cathode, to a surface where they are deposited as atoms of plated metal. In electroforming, the receiving surface is called a mandrel; it is conditioned so that the plating does not adhere. The plating can be lifted away, retaining its deposited shape, as a discrete component.

This process endows parts with the following unusual characteristics.

1. They can have very thin walls, down to 0.001″ thick.
2. Surface features of the mandrel are reproduced with extreme fidelity and high surface finish.
3. Complex contours are no more difficult to reproduce than the simplest shapes.
4. Some dimensional tolerances can easily be held to 0.0005″.
5. The available plating tanks are the only limit to the size of parts to be electroformed.

MANDRELS

The same relationship exists between mandrel and electroform that exists between a mold and its casting, i.e., internal features of the electroform are formed as negative images of the mandrel.

Both permanent and disposable mandrels are used. When there are no undercut surfaces on the electroform and it can be lifted directly off the mandrel, a permanent mandrel is generally used. If undercuts are necessary on the electroform, the mandrels must be destroyed by dissolving or some other device. (See Fig. 8.1.)

Stainless steel and aluminum are the materials most often used for mandrels. Stainless polishes easily and provides internal finishes as high as 2 microinches. External finishes are normally about the same as those of a die casting. Aluminum mandrels machine much more easily but do not have the service life of stainless steel. There are several other materials used for mandrels: bismuth alloys, invar, and cast alloys of many different metals like brass or nickel. Plastics, wood, wax, and quartz can also be used. Nonconductors must first get painted with a conductive material before use as mandrels; and since even the smallest scratch will reproduce, mandrels must be carefully handled.

When large numbers of disposable mandrels are required, castable or moldable units are used if at all possible. There are a number of tradeoffs to be considered when contemplating mandrels. Because of the quantity of mandrel materials available, and the special features of many of them, this becomes a serious decision.

The necessity for undercuts should not be combined with a

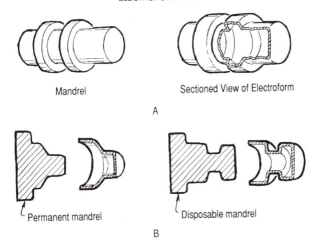

Fig. 8.1. Mandrels used in electroforming. **A. Shape and Finish:** part is formed by electroplating over a removable or disposable mandrel. Internal surface desired on the part is machined or formed as "negative" external surface on the mandrel. **B. Mandrels:** mandrel may be reused if the part can be removed directly. A disposable mandrel, which is destroyed as the part is removed, must be used if the part contains undercuts. (Courtesy of Gar Electroform Division.)

requirement for high finish or close dimensions. If such requirements are unavoidable, more than one material can be used for the mandrel by bonding. This technique is often used for waveguides.

Although most mandrels are "male," sometimes the need for a fine external finish or a tight external dimension requires the use of a "female" mandrel.

WALL THICKNESS

Electroforms are usually parts of constant wall thickness. The thickness may vary from 0.005" to 0.5", although most electroforms are in the range of 0.01"–0.05" thick. Uniformity of thickness is subject to the same vagaries as in electroplating, i.e., areas of high plating current will thicken the deposit locally as will sharp edges or convex surfaces.

The simplest way to avoid these thickness variations is to provide adequate radii at edges and corners, and to make holes and slots as wide as they are deep; otherwise, the plater must use shields and "thieves" to reduce current density, or he can use conforming anodes to boost current density. (See Fig. 8.2.)

INSERTS AND GROW-ONS

The desired part need not be made entirely from deposited metal. Other materials, even nonconductors, may be incorporated into the component by plating over or around separate

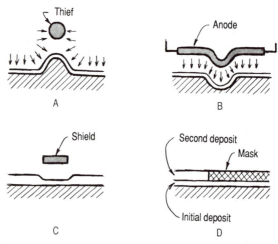

Fig. 8.2. Methods used to vary wall thickness in electroplating. **A. Providing Uniform Wall Thickness:** cathodic "thief" reduces current density to prevent excessive plated deposit on high spots. **B.** Conforming anode provides required current density within recesses. **C. Varying Wall Thickness:** shield decreases current density to provide gradual variation in wall thickness. **D.** Mask provides abrupt change in wall thickness. (Courtesy of Gar Electroform Division.)

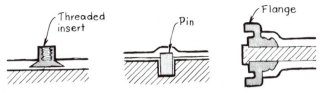

Fig. 8.3. **Inserts and Grow-ons:** plated metal can be deposited around separate pieces, called grow-ons, attached to the mandrel. Grow-ons then become an integral part of the electroform. (Courtesy of Gar Electroform Division.)

pieces attached to the mandrel. Threaded inserts or shafts or other "grow-ons" are often used in this manner.

Sometimes the nondeposited metal constitutes a larger portion than the electroformed material. The joining of two or more pieces is now used for a large percent of all electroformed output. (See Fig. 8.3.)

MATERIALS AND PROPERTIES

At this time, nickel is the material most often electroformed. It provides good strength and corrosion resistance, and it is an easily plated metal. Copper is the next most popular material. Gold, silver, and rhodium are used when extreme resistance to oxidation is the prime characteristic desired.

Before progressing too far in the design, the assistance of an electroformer should be sought because the actual plating conditions, such as composition of the plating solution, has a lot to do with the results. Properties of deposited metal can vary considerably because of changes in the plating bath. Residual stresses in an electroform may occasionally cause difficulties for highly stressed parts; and proper heat treatment can make quite a difference in properties.

Sometimes, one metal will be deposited on top of another when the properties of the first are not good enough for the task. By manipulating with masks and shields, many combinations of shapes and functions can be achieved. Chromium, for instance, is too brittle for electroforming; but if deposited with nickel, the combination is usable.

Electroless nickel (92% nickel and 8% phosphorus) deposited on electroformed copper can be heat treated to a hardness of 70 Rc, and many such combinations are used. Research and development in electroforming have led to the addition of ceramics to certain plating baths; these additives become embedded in the electroform as the plated metal is deposited, substantially increasing strength and resistance to creep.

One feature of an electroform establishment is the unusual cleanliness in evidence. In order to ensure clean and flawless parts, free from imperfections caused by dust contamination, the interior air must continuously be filtered. All production electroforming baths have unitized filters for each tank. Physical characteristics of the desired electroforms are maintained by conformance to specialized formulations.

9

ELECTRICAL DISCHARGE MACHINING

The electrical discharge machining (EDM) phenomenon was first noticed around 1700. However, it was not until around 1948 that the Russians first applied the principle to metal stock removal. The popularity of this machining method has grown by leaps and bounds in the last 30 years; lately, its growth rate has been about 30% annually.

Machine power, speed of stock removal, and types of jobs EDM can do better than any other machining method have increased to the point that many jobs must now be done by EDM (conventional or wire) in order to be competitive.

PROCESS

EDM is a precision metal removal process using an accurately controlled electrical discharge (spark) to erode metal. This process will machine any electrically conductive metal, regardless of its hardness.

To visualize the process, picture one electric spark passing from a negatively charged ($-$) electrode to a positively charged ($+$) workpiece, both of them immersed in a bath of dielectric oil. The energy of the spark brings particles of the workpiece to a vaporized state. These particles immediately resolidify into small spheres and are flushed away by the dielectric oil, leaving a small pocket eroded in the workpiece. This cycle, repeated thousands of times each second, erodes material from the workpiece until a reverse image of the electrode is formed in the workpiece. (See Fig. 9.1.)

A good analogy is the comparison of EDM to a thunderstorm. The thunderstorm has a negatively charged cloud, a positively charged cloud, and wind movement. The EDM circuit

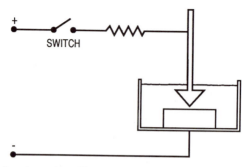

Fig. 9.1. An elementary EDM machine. Shown here: a workpiece sitting in a tank of dielectric oil, the electrode, a source of electric power connected through a switch, and a servo mechanism to advance the electrode into the eroded crater in the workpiece.

has an anode, a cathode, and a switch with a servo system. The air gap between the two clouds acts as resistance similar to the dielectric oil of EDM. As the wind blows the clouds toward each other, the potential energy overcomes the air gap resistance and electrons will jump the space. The electrons that ionize the gap seeking unbalanced atoms generate the tremendous energy (lightning) which causes havoc. EDM uses the same principle on a controlled basis to vaporize conductive metal.

There was a simple evolution to EDM. It began with an RC circuit (resister–condenser) which provided random discharges — some small, some large. This combination resulted in irregular surface finishes, slow cutting, different overcuts, and excessive electrode erosion.

The present EDM process is used for drilling small holes to machining huge 50-ton dies. The advent of CNC wire-cut systems has dramatically expanded both the quantity and sophistication of EDM applications, which have improved the versatility and profitability of tool making operations. In fact, they have also made EDM the logical choice for many production activities as well.

If we substitute a transistor for the switch, the on–off function can be achieved thousands of times per second. Another advantage of the transistor is that there is no buildup prior to turning on. The new type of power supplies have further improved control by assigning independent values to "on" time and "off" time. The "on" time creates the spark crater, while the amount of electrical energy determines the crater size.

The EDM system is composed of two components — a machine tool and a power supply. The machine tool positions the electrode in relation to a workpiece in order to erode a cavity of some type. The power supply produces a high-frequency series of electrical arc discharges between the electrode and the workpiece which erodes metal from the workpiece.

Fig. 9.2 illustrates the components of an EDM system. The electrode is attached to the ram of the machine tool. A hydraulic cylinder or DC servo unit moves the ram (and electrode) in a vertical plane to position the electrode in relation to the workpiece. This positioning is done automatically by the servo driven by the power supply. During normal machining, the elec-

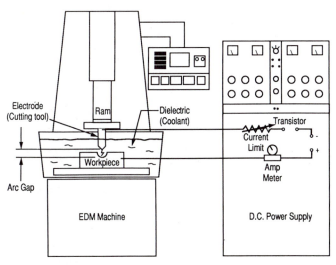

Fig. 9.2. The EDM system. Electrode and workpiece are held by the machine tool (left), which also contains the dielectric system. The power supply (right) controls the electrical discharges and the movement of the electrode in relation to the workpiece. (Courtesy of Elox Corporation.)

trode does not touch the workpiece because they are separated by a small gap.

Both the electrode and the workpiece are immersed in dielectric oil. The same oil acts as a coolant, and is pumped through the gap to flush away eroded particles (swarf). In operation, the ram moves the electrode toward the workpiece until the space between them is such that the voltage in the gap can ionize the dielectric and allow a discharge to occur.

During the off-time, the oil regains its insulating properties. It remains in this state until reionized by the next pulse. This process is repeated thousands of cycles per second. Each discharge melts a small area of the workpiece. The molten metal cools, solidifies, and is washed away by the flushing action of the dielectric oil.

Note that the hardness of the material has no effect on the process. The material could be hardened tool steel or even carbide. Rather than machine a part before heat treating it, EDM permits the machining to be done after hardening; this eliminates risk of distortion or any other damage. Graphite, copper, or tungsten are generally used to make electrodes. The electrode is always made slightly smaller than the cavity desired because the erosive action progressing outward from the electrode always produces a cavity slightly larger than the electrode. This size difference is called overcut. Once established, overcut is predictable.

POWER SUPPLY

The power supply controls the amount of energy consumed. First, it has a time control function which controls the length of time that current flows during each pulse; this is called "on time." Then it controls the amount of current allowed to flow during each pulse. These pulses are of very short duration and are measured in microseconds. There is a handy rule of thumb to determine the amount of current a particular size of electrode should use: for an efficient removal rate, each square inch of electrode calls for 50 amps. Low current levels for large electrodes will extend overall machine time unnecessarily. Conversely, too heavy a current load can damage the workpiece or electrode.

The impact of each pulse is confined to a small area, the

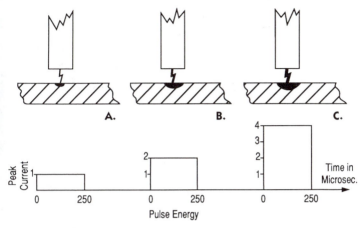

Fig. 9.3. Metal removal rates increase with the amount of energy
per spark. (Courtesy of Elox Corporation.)

location of which is determined by the shape and position of the
electrode. The arc always travels the shortest distance, across
the narrowest gap, to the closest "high spot" on the workpiece.
After each "high spot" is removed, the impact goes to the next
highest, and so on until the machining is complete.

Fig. 9.3 shows how metal removal increases as energy input
increases. The energy depends upon the number of sparks per
second and the amount of energy in each spark; this energy is
measured in amperes. In Fig 9.3, view B shows the removal of
twice the metal taken from view A; that is because twice the
energy was used. This is repeated in view C: twice the energy in
view C removes twice the metal that was removed in view B.
The amount of metal removed is normally proportionate to the
energy used.

Surface finish is important on many jobs, and it is a function
of two things: "on time" and peak current, which are both settings of the power supply. Another rule of thumb tells us that
long "on time" and/or peak current produces a rough finish,
and that short "on time" and/or low peak current produces a
fine finish as shown in Fig. 9.4.

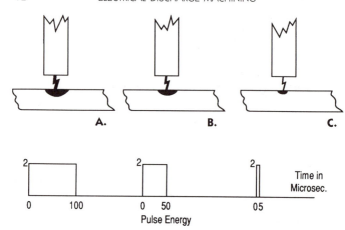

Fig. 9.4. Surface finish improves as the spark energy decreases. Energy at the spark may be increased or decreased by changing the peak current and/or on-time. (Courtesy of Elox Corporation.)

USES OF THE PROCESS

Difficult shapes, in general, are machined as easily as simple shapes by EDM. The electrode of the complex shape is more difficult to prepare, but the machining is no more difficult. The following are readily machined: deep narrow pockets, sharp inside corners, thin ribs, twisted shapes, round corners, multiple indentations, narrow slots, and various-sized apertures. Blind cavities and through holes demand different techniques, but both can be handled by EDM.

All types of tools, dies, and molds can be made by EDM, saving time and money. Injection molds for plastics, metal injection molds, and other plastic molds, die casting, cold heading, coining, forging, stamping, extrusion, and powder metallurgy dies can all be made by EDM less expensively than by using the old type of machine tooling approach. In fact, there are situations when the use of wire EDM can produce male and female die components with one cut (more about this later).

During the EDM process, there is no need for heavy

mechanical clamping; clamping can be light and delicate so that fragile parts can be machined without distortion.

Surfaces which have been EDM'd have a nondirectional texture; they will be covered with tiny pockets like a shot-blasted surface. When fine finishes are required (better than 100 RMS), it is better to polish by hand. Fine finishing can generally be done by EDM using a tungsten carbide electrode, but that would slow down the process.

EDM can machine to a tolerance of 0.001″ consistently, which includes the repeatability required when going from one electrode to another. The major roadblock to accuracy is the electrode: if the electrode is not accurate, the workpiece will reflect the inaccuracy.

ELECTRODES

One of the first decisions the EDM user must make concerns the choice of electrode material. With the exception of tungsten and graphite, most electrode material melts near 1000°C, which means they will lose material during the process. Graphite sublimates (changes directly from a solid to a vapor) at about 3350°C, so the EDM process will not erode graphite as quickly as metal.

An electrode's ability to resist wear has a significant effect on the surface finish of a machined part. An important feature of electrode material selection is machinability. Except for tungsten, all the common electrode materials machine fairly easily, but graphite combines good machinability with good EDM performance.

The cost of making electrodes can sometimes be discouraging. There are at least two machines on the market which eliminate this problem. At one company — Easco-Sparcatron — they have developed the "Total Form Machine." This is, in effect, a carbon copier, which turns out copies from a master. This machine produces copies in minutes rather than hours compared to conventional machining. At least one other electrode manufacturing machine is on the market.

Electrode material is usually selected according to:

1. metal removal rate,
2. resistance to wear,
3. surface finish obtainable,

4. expense,
5. machinability.

Electrode wear is exhibited in four ways:

1. corner wear,
2. end wear,
3. side wear,
4. volumetric wear.

The most significant of these is corner wear, since it determines the accuracy of the final cut contour.

FLUSHING

The single most important factor in successful EDM work is flushing: the removal of metal particles from the gap. Good flushing provides good machining conditions. If the flushing is poor, metal will be removed erratically, time to machine will increase, and costs will increase. Sometimes, when the overcut is 0.0015″ or smaller, a vacuum flush will be used. Sometimes a combination of methods will be used.

Oil which appears black in the workpan is not necessarily dirty in the gap. The oil in the gap comes directly from the filter, while the oil in the pan has been used. Most machines have an adequate oil circulation system. Several companies market dielectric oil especially for EDM work.

Again, flushing swarf (eroded particles) from the cavity is one of the most important features of efficient EDM machining. The electric spark does not differentiate between particles and the workpiece material and thus will react upon the closest oppositely charged particle. The spark gap between the electrode and particle being less than the predetermined optimum gap will cause the servo to respond, thus removing the electrode from the cavity. This situation is also conducive to the creation of a hotspot or arc.

Fig. 9.5 shows the customary methods of flushing swarf: internal pressure, vacuum (suction) patterns, and external dielectric flow. It should be remembered that volume of dielectric flush is more efficient than high pressure. The use of pressure flush can cause turbulence in the cavity and coagulation in areas of conflicting flow patterns.

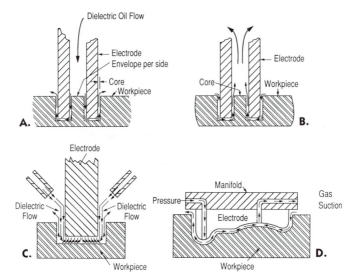

Fig. 9.5. Examples of flushing methods. **A.** Pressure flush through electrode. **B.** Suction flush through electrode. **C.** External flush. **D.** Combination pressure and suction flush. (Courtesy of Easco-Sparcatron.)

SMALL HOLE DRILLING

This subject deserves special attention. Many parts in instrumentation, fuel injectors, and turbine blades demand very small holes. Such holes, 0.005″–0.015″ in diameter, can be drilled in hardened steel by EDM. Traditionally, this work was done by a skilled machinist using a sensitive drill. Broken drills and scrapped parts made this an expensive operation.

EDM equipment in general use is not able to perform such work without serious problems such as recast layers (surface EDMed, melted, and reformed) and spatter; that is because most EDM work is large scale (e.g., die jobs) and small hole drilling requires a different technique. However, it is possible to make and use special low power equipment as the Raycon Company of Ann Arbor, MI, has done. With proper controls, accuracy and repeatability of small hole EDMing can be accomplished successfully and inexpensively. An operator could do small hole EDMing instead of a skilled machinist using a conventional drill.

WIRE EDM

This is another specialty development of the basic EDM. Today, many molds and dies are produced by wire EDM because this technique saves time and money. In wire EDM, a traveling, thin wire electrode cuts through a workpiece as it discharges. The wire is started through a hole (which has been drilled or EDMed) and then it is fed by a microprocessor to cut out any shape required. The wire, generally brass, is held in top and bottom guides and moves through the workpiece like a bandsaw. (See Fig. 9.6.)

Since it was introduced around 1970, wire EDM has increased in popularity much more quickly than any other machine tool. Its feed rates have increased from 1.5 cubic inches per hour to about 23.

The wire eliminates the need for elaborate and precision electrodes which are necessary for conventional EDM work. The precision is in the memory of the CPU, and a new die or tool can be made a month or a year later because the die's shape is always available in memory. A computer numerically controls the wire (electrode) movement so the workpiece is made with exceptional accuracy. As power supply technology has improved, surface finishes have also improved.

Punch and dies can be machined by wire, holding tolerances well within 0.001″. One cut with wire, selected from a range of 0.006″−0.014″ diameter, can produce both punch and die from one block of tool steel (see Fig. 9.7). Generally, roughing cuts are within 0.004″ of the finish size. When the slug is removed, and power levels reduced, it normally takes 2 or 3 finishing cuts to get the desired dimensions and the proper finish. The total time for finishing cuts is about twice the time for roughing. Turn-around times are fast, and finishes of 10 RMS are commonplace.

Wire EDM is similar to die sinking EDM, which we have been calling conventional. Die sinking uses a plunging electrode, while wire EDM uses its traveling wire. But each can be used in place of the other.

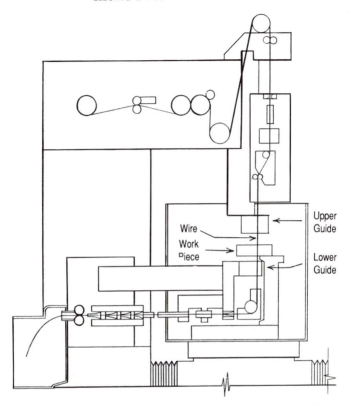

Fig. 9.6. Typical wire EDM machine. (Courtesy of Jatco Machine Company.)

Conductive ceramic can be cut by die sinking EDM, but one must be certain that the ceramic is truly conductive. If it were not conductive at some spot, the servo might interpret the lack of conductivity as an open circuit condition and move the electrode into the workpiece causing a crash. This kind of situation is less significant for wire EDM because in that case the wire would simply break.

In conclusion, wire EDM is a versatile machine that has a broad range of applications.

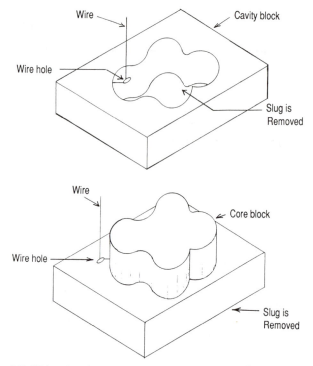

Fig. 9.7. With wire EDM, one cut produces both punch and die. (Courtesy of Jatco Machine Company.)

GLOSSARY OF EDM TERMINOLOGY

Amperes—The unit of measurement of electrical current.

Arc gap—The distance between the workpiece and the electrode when EDMing.

Capacitor—A component that stores a charge.

Coolant—Same as dielectric.

Cycle—Made up of the on time and off time, and expressed in microseconds, it is the time associated with the discharge of one EDM spark.

Dielectric — The oil used in an EDM system as a coolant to flush away eroded metal particles, and used as a barrier between the electrode and the workpiece.

Electrode — It is a formed part made of electrically conductive material, used to machine the workpiece; it can be male or female in shape.

Flushing — Forcing dielectric through the arc gap to remove eroded particles.

Frequency — The number of sparks per second.

Gap current — The number of amperes flowing between the workpiece and electrode; it can be read on the ammeter.

Gap voltage — The changing voltage, monitored on the voltmeter, which appears across the electrode–workpiece gap.

Ionization — The phenomenon by which dielectric oil breaks down and becomes electrically conductive.

Machine tool — The mechanical system which holds and positions the workpiece and guides the electrode.

Microsecond — Equal to one-millionth of a second, it is the time unit used to measure on and off times in the EDM cycle.

Off time — The time of cycle during which no current flows; this time allows the molten metal to resolidify and get flushed from the gap.

On time — The time during which current flows, expressed in microseconds.

Overcut — The clearance per side between the electrode and the workpiece after the machining operation.

Peak current — The amount of current which flows during on time.

Power supply — The electronic system which generates and controls the electrical discharges.

Servo — The system which converts electrical signals into mechanical motion to move the electrode.

Short circuit — This occurs when the electrode and the workpiece are in direct contact; it causes the machine to stop and the ram to retract.

10

ABRASIVE WATERJET CUTTING

PROCESS DESCRIPTION

Waterjets cut material by a compressive, shearing action. The jet reaches velocities of close to 3000 feet per second after being pressurized up to 55,000 psi—a velocity about 3 times that of a pistol bullet. Any soft material in its path will simply be removed.

A.

Fig. 10.1. **A.** Five-eighths-inch thick Kevlar being cut with AWJC. (Courtesy of Ingersoll-Rand Corp.) **B.** Stacked materials being cut simultaneously. From top, the materials are: ½" thick aluminum; ⅝" thick Kevlar; ½" plate glass; and ½" phenolic. (Courtesy of Ingersoll-Rand Corp.) **C.** A graphite epoxy block that has been cut by an abrasive waterjet. (Courtesy of Textron Aerostructures/ Textron Inc.)

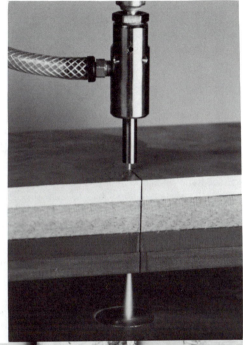

B.

C.

Harder materials, such as glass, metals, and ceramic, are cut by adding an abrasive to the jet; thus, the waterjet drives the abrasive and is not the primary force any longer (silica and garnet are the abrasives most usually used). Currently, the materials being cut by abrasive waterjets include aluminum, steels, titanium, superalloys, glass, concrete, and tough composites. Cutting speeds for 1″ thick pieces of these materials range from 2″ to 18″ per minute.

Three major components make up a waterjet cutting system: the intensifier, nozzle, and receiver. The intensifier houses a pump which generates water pressures of 30,000–55,000 psi. The receiver catches the blast of abrasive-laden water and directs it to a sorting tank. The nozzle is described below.

The Ingersoll-Rand intensifier unit is available, computer controlled, in enough sizes and capacities to suit almost any requirement. It can be used with a variety of machines for a large number of job types. For instance, it may be used with a handheld cutter (Fig. 10.2), which is an articulating arm and balancing device which supports the manually guided WJC system. It could also be used with a waterjet slitter, which performs con-

Fig. 10.2. Ingersoll-Rand CNC waterjet cutting system.

tinuous slitting of web material (multiple nozzles are position adjustable for maximum flexibility). Another use could be with a manual cutting station, which performs free hand or by following a template (a tilt table is optional), or with the CNC waterjet cutting system, which provides accurate, reliable solutions to both waterjet and abrasive waterjet operations. This machine is for different work than handled by the machine shown in Fig. 10.2.

Other specialized cutters used with the intensifier are the circuit board cutter, which has dual fixtures for maximum circuit board production, and the shape cutter, which is used for hydrobrasive contour cutting of repetitive shapes of various materials. Different patterns may be cut by substituting profile templates. It could also be used for the cable stripper which automatically strips and cuts cable insulation without damaging the metal conductors.

Ingersoll-Rand, the company which created this intensifier for use in both waterjet and abrasive waterjet (hydrobrasive) operations, makes three pumps: the Streamline Intensifier 25S, 40S, and 50S. With proper selection of intensifier capacity, a single intensifier can supply high pressure water for multiple applications in separate locations.

The design pressure for these pumps is 60,000 psi. The maximum operating pressure is 55,000 psi. These are high ratio, slow stroking, and double acting, and each has a calibrated rupture disc for safe venting of overpressures.

An extensive range of waterjet cutting systems are available from 25–100 HP, for single or multiple cutting applications. Motion control equipment from manual cutting to multiaxis CNC profiling is also available.

The state-of-the-art Ingersoll-Rand computer-controlled water intensifiers come in enough sizes and capacities to suit most any requirement, and they can operate the company's standard cutting machines.

It may be difficult to believe, but abrasive waterjets can cut through 14″ of concrete and 3″ of steel plate at rates of 1.5″ per minute in a single pass. Equally impressive is the fact that this cutting produces no heat which can degrade metallurgical properties (plasma arcs can usually cut materials a little faster than waterjets, but will leave a heat-affected zone). The usual cut edge leaves a surface roughness of 150–250 microinches.

Around 1970, the author used a similar process to clean metal surfaces before adhesive bonding; that was around the time that Ingersoll-Rand began producing commercial systems. In 1985, Dr. Gerin Sylvia, Professor of Industrial Engineering at the University of Rhode Island, started testing an automated abrasive waterjet cutting system. He cut everything from 2″ thick armor plate, boron reinforced aluminum, gray cast iron to 2″ thick titanium.

One of the advantages of abrasive waterjet cutting (AWJC) is its easy maneuverability. Bridge inspection is an area where the technique is valuable. AWJC can cut through thick slabs of reinforced concrete and expose the underlying steel framework. The framework can then be examined easily, without the mess and labor of jackhammers and saws.

In the foundry, AWJC is being used to cut off gates and risers from castings. When castings are stripped this way, they should be dried to avoid corrosion. (Operators might have to wear hearing protection when using this method.) Various abrasives work faster than foundry sand, but the sand is so inexpensive and readily available that most foundries use it instead of the customary abrasive.

Nozzles with a small ID make a deeper and higher quality cut than those with a large ID. Nozzle life has increased to between 4–8 hours of constant use. At the present time, a good nozzle can cut 0.06″ thickness of gray cast iron at 36″ per minute, and 0.88″ thickness at 4″ per minute. A hydrobrasive nozzle is being used by North American Aircraft to cut components for the B-1B bomber. Parts for the bomber are cut from titanium sheet 0.125″ thick. In this case, the resultant kerf is 0.075″ wide, but if the material is thicker, the kerf will vary in width from top to bottom. The parts generally are burr free. Operating pressure is 50,000 psi using 2 pounds per minute of #60 grit red garnet abrasive and 1 gallon per minute of water. The garnet is introduced into the mixing chamber by vacuum caused by the moving jet. This is the same technique used for mixing weed killer or fertilizer into a garden hose waterjet.

One of the main reasons for using AWJC on the bomber's titanium sheets is the fact that the cuts are almost burr free. This allows a ratio of better than 10 to 1 in favor of AWJC over the conventional cutting method in producing burr free titanium pieces.

In 1989, Lockheed Aeronautical Systems Co. of Georgia began using an AWJC system which combined abrasive waterjet technology and computer aided design. The system was originally used to cut production parts for the C-17 and the C1-130 Hercules. The waterjet workstation uses an Ingersoll-Rand cutting nozzle and a Cimcorp controller. The operator secures the material to be cut on a work table, then the nozzle is automatically guided. The guidance is achieved by the operator selecting instructions from the computer menu, and a 6-axis robot-controlled waterjet delivers a mixture of abrasive material and water through the nozzle.

The main disadvantage of AWJC is the short life of nozzles, although nozzle life has already been increased from 20 minutes to 4 hours, and improvements are expected to continue. Improvements to separate solids from spent water are also being worked on. In spite of these problems, many users of AWJC report large savings.

The new materials currently used in aircraft, such as composites, have posed fabrication problems. It is difficult to cut them without causing delamination and ragged edges. AWJC is filling this need. In one case, an aircraft door made of graphite epoxy required 8 hours of machining; the job is now being done by AWJC in 12 minutes.

At present, the nozzle is being controlled either manually or by machine, and sometimes by computer. The manual method is not as easy as it may sound. The operator requires considerable skill to position the wand around the workpiece at the proper cutting speed while holding the wand the correct distance from the work. Flow Systems Corp., Kent, WA, is furnishing turnkey AWJC systems involving automation, robots, and machines.

High pressure cutting produces a reaction force of approximately 30 psi. While this is considerable to handle manually, it is very little for a machine to control. The development of rotary joints and high pressure water lines have done a great deal to advance the application of AWJC.

The third major component of the waterjet cutting system — the receiver — is undergoing continued research. After the abrasive jet has cut through the workpiece, it is still traveling at supersonic speeds. Its potential to erode anything in its path requires some type of receiver to avoid trouble; sacrificial tooling is generally used for that. Research is continuing to develop better receiving systems. A catch tube, 2–4 feet long, lined with carbide is most commonly used at present.

The initial cost of AWJC systems is usually competitive with the cost of lasers. Apparently, lasers are most cost effective at cutting materials less than 0.375″ thick. Power requirements for lasers cutting thicker materials become too costly. Very often, an application which is not right for lasers can be handled nicely by AWJC and vice versa.

11

MAGNEFORM

HOW MAGNEFORM WORKS

The basic magnetic pulse principle has been in use in metal forming equipment for about 22 years. It is the same principle at work when an electric motor is activated. Magneforming is when an electric current generates a pulsed magnetic field close to a metal conductor so that a controllable force is created which can be used to shape metal without actual contact. (See Fig. 11.1)

The basic components of the Magneform machine are: energy storage capacitors, a work coil, and switching devices. High voltage capacitors are charged, then discharged through a coil, inducing an intense magnetic field. This field, in turn, in-

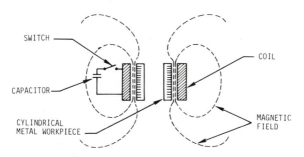

Fig. 11.1. In the basic Magneform circuit, high voltage capacitors are discharged through a coil, inducing an extremely intense magnetic field. This field in turn induces current in the conducting workpiece, setting up an opposing magnetic field. The net magnetic force does the forming. (Courtesy of Maxwell Laboratories, Inc.)

duces current in the conducting workpiece setting up an opposing magnetic field; the net magnetic force does the forming.

During forming, pressures as high as 50,000 psi move the
workpiece at velocities as great as 900 feet per second. The
strength of this force can be closely controlled, which is important for versatility. These magnetic forces can produce up to
180 parts per minute.

Magneform is a proven, widely accepted method of forming
metal parts. Controlled forces are placed where they are wanted;
and the machine cycle can be easily synchronized with con-
veyors and other feed mechanisms. The actual forming can be
done in 100 millionths of a second.

The following are past work assignments.

Applications	Production Rate/Hour	Mode of Operation
Ball Joint Seals	2400	Automated
Automative Fuel Pumps	900	Automated
Timer Rotors	720	Automated
Baseball Bats	450	Semiautomated
Projectile Rotating Bands	400	Manual

BENEFITS OF MAGNEFORMING OVER CONVENTIONAL FORMING

Magneform enjoys the following advantages.

1. It is precisely controllable and allows forming of metal
   parts over plastics, composites, glass, and other metals.
2. Since Magneform makes no contact with the workpiece,
   no lubricant is required. Parts have no tool marks, heat
   deformation, or surface contamination.
3. Magneform parts are uniform in appearance and there is
   no tool wear.
4. Magneform makes joints equal to or stronger than the
   material of the workpiece.
5. Magneform machine operators are easily trained.
6. The machines are energy efficient and easily installed.

No conventional equipment can duplicate Magneform's ability to concentrate a uniform force to a selected area on the

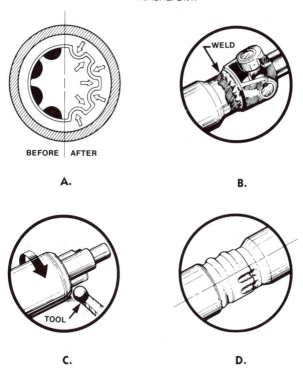

Fig. 11.2. Magneform joining, as seen in "**A**" and "**D**," has many advantages over welding ("**B**") or spinning ("**C**"). (Courtesy of Maxwell Laboratories Inc.)

periphery of a workpiece without mechanical contact. Two parts can be joined concentrically and without excessive heat. Even if one of the parts has eccentric protuberances, the two parts can be joined concentrically. Fig. 11.2A indicates how an outer shell will fill in serrations. Although the parts in this sketch are concentric, if the inner part had eccentric protuberances, the outer part would deform completely around the inner.

Fig. 11.2B shows a welded joint which would be neater and cheaper if it were Magneformed. Problems which occur with welded or brazed joints are: changes in physical properties in metal surrounding the weld, spatter requiring removal, special care removing flux, and a rejection rate especially when seals are made.

Fig. 11.2C shows a spun joint which is a type often used when heat cannot be applied. Spinning requires a skilled operator to assure minimum eccentricity. This is another place to use Magneforming, which would do the work faster and simpler. A low-skill operator could handle the work; and even if the parts were out of round, Magneforming would join them with little difficulty.

The joint of Fig. 11.2D indicates a typical, neat Magneformed job. This speeds up assembly and makes automation possible.

Magneform also has advantages over a pinned or dowelled assembly, which requires drilling then assembling pins, rivets, bolts, or dowels, thus creating localized stress.

Magneforming can simplify and improve production procedures in several ways. The inherent advantages of this method preclude certain problems common to the conventional joining methods.

MAGNEFORM APPLICATIONS

Applications include forming and assembly operations in the automotive, appliance, aerospace, electrical, nuclear, and ordnance fields. Magneform generally replaces swaging, spin rolling, soldering, pinning, or welding operations. There are some jobs which cannot be done at all by any other method.

Electromagnetic metal forming works best with light to medium gauge materials which have high electrical conductivity such as copper, aluminum, steel, and brass. Stainless steel and other materials with poor conductivity can be formed by using a high conductivity driver such as an aluminum or copper sleeve. The workpiece must provide a continuous electrical path.

The magnetic work coil may be located remotely from the energy storage and control unit to accommodate any production requirement. The work station can be designed to include automatic workpiece orientation and processing.

FORMING PRESSURE TIGHT TORQUE JOINTS WITH MAGNEFORM

Pressure tight joints, capable of withstanding high wrenching torque, can be quickly assembled by Magneform. Considerable money can be saved by using this method rather than welding or brazing. Also, the appearance of the joint will be improved.

A typical assembly made with a single magnetic impulse is shown in Fig. 11.3. The tube is 3003-h14 aluminum with an O.D. of 4.8"; the cap is 6061-t6 aluminum. The requirements are as follows: it must be leak tight to 400 psi and must withstand a wrenching torque of 140 ft-lbs and a tensile load of 7000 lbs.

This design employs an O-ring for a sealant and a coarse, straight knurl for torque resistance. Test results indicated that the assembly exceeded all requirements. When considering the use of magnetic impulse forming in this type of assembly, the groove configuration must be tailored to the function. This will vary depending on the material and the operating environment (as might prohibit the use of a rubber O-ring). The addition of a knurl increases the joint strength in torque. When thin wall tube material is to be used, a band can be formed around the assembly.

To eliminate the contamination caused during manufacture of high performance fluid system components, Michigan Dynamics Inc. devised an unusual method for producing ultra-clean 10 micron wire cloth filter assemblies for the Sidewinder missile launcher. The filter components are sealed in plastic bags in a clean room after first being decontaminated and semiassembled.

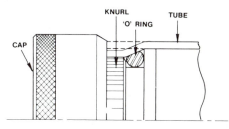

Fig. 11.3. Schematic showing configuration of typical pressure tight torque joint.

These filter components are easily cleaned in the clean room prior to assembly. Still in the clean room, the parts are placed and held in the proper relationship by the snug fit between components. Next, the loosely assembled parts are sealed in a polyethylene bag. Then the packaged parts are taken out of the clean room to a Magneform machine in the manufacturing area without the possibility of contamination. The bagged parts are placed in a magnetic field shaper which concentrates a magnetic field around the area to be swaged.

CONTOURED TUBING TRANSITIONS

The round-to-square transition is accomplished with a single magnetic impulse from a Magneform expansion coil. The unformed tubing is placed in a split die with the Magneform coil inside the tubing in a position where the maximum field intensity is exerted in the area where the transition is to be formed. Transitions of this type can be readily formed at the ends of any lengths of tubing in diameters of from approximately 1.5" and up, and in wall thicknesses of from approximately 0.020 to 0.080".

PRODUCTION COSTS OF DUCTING INTERSECTIONS

Tubing parts used for ducting intersections are employed in a wide variety of diameters, contours, and intersecting angles. Most of these parts were formerly produced as welded joints. Using Magneform expansion coils and female dies, the parts are now formed in a single operation with significant cost savings.

MAGNEFORMED JOINTS COMPARED TO OTHERS

Magneform swaging is entirely a machine operation. Once the parameters of the operation are established, and the joint is positioned and aligned by fixtures, the only step which remains to be taken is the release of energy by actuating a switch. By contrast, welding requires highly specialized operator skill. The strength of the joint is variable and depends on the skill of the craftsman, whereas the output of the Magneform process is more reliable and consistent. Sometimes high strength and cost are secondary to reliability. Consistency of strength is the greatest asset of the Magneform process. Since this process doesn't use heat, expansion, shrinkage, and bowing problems are avoided.

COMPARISON OF MAGNEFORM AND WELDED JOINTS

Conditions	Welded Joints	Magneform Joints
Load at failure	5680 lbs	5377 lbs
Ultimate tensile stress	12,000 psi	11,400 psi
Variation in samples (%)	17.8	13.97
Corrosion properties	Good	Good
Resistance to vibration	Good	Good
Strength uniformity	Fair	Excellent
Inspection	Radiography and dye check	Visual

Fig. 11.4. A steel cover is formed over a cast aluminum body to assemble this mechanical fuel pump. Cycle time is 10 seconds. (Courtesy of Maxwell Laboratories, Inc.)

No conventional equipment can duplicate Magneform's ability to join two parts concentrically, without heat or mechanical distortion, nor as quickly. Welding distorts and embrittles; brazing distorts metal, and brazing aluminum takes a fair amount of skill; spinning joints requires a skilled operator; bolting concentrates the load in a joint. Magneform does not embrittle or distort metal, nor does it spatter. It speeds up assemblies and should be the selected method for this type of work. (See Fig. 11.4 for an example.)

12

FINEBLANKING

BACKGROUND

Fineblanking was invented in Switzerland in the 1930's, but it has been very slow to catch on. It cannot take the place of conventional stamping because it is a much more expensive machine and is much slower in operation. It simply cannot compete with most blanking or stamping jobs. However, for those applications where edge quality, flatness, and accurate dimensioning are required, fineblanking may be the best method to use because it saves one or more secondary operations.

For years, the extra cost of fineblanking machines and tools just didn't seem worthwhile. If a comparison is made of only strokes per minute, stamping press wins easily. Also, this newer process requires careful attention to detail. Nevertheless, the main reason fineblanking was overlooked was that it was unknown or at least little understood. But the picture is beginning to change. Industry is becoming receptive to anything that reduces cost of manufacturing. At present, there are three foreign companies selling fineblanking machines in the United States: Feintool, Schmid, and Hydrel; and both Feintool and Schmid now assemble machines in this country.

A trend that is boosting fineblanking is the growth of statistical quality control programs. Fineblanking complements this movement because it is a technique oriented to quality. Quality is built into the machine and the tools. In addition, quality is easier to monitor when parts are produced on one high quality machine instead of several machines using different processes.

PROCESS

Fineblanking uses special presses and special tools to stamp metals so that the parts are sheared over the full material thickness in one stroke. In most cases, only belt sanding or barrel tumbling is necessary to finish the part. One major difference between fineblanking and stamping is the clearance between punch and die. In stamping, 5–10% of the material thickness is the usual clearance. In fineblanking, the clearance is 1% or less. In fineblanking, a V-ring follows the outer shape of the blanked workpiece. When the tool closes, this raised V-shaped ridge bites the stock material, holding it tightly so that it cannot flow away from the punch during the blanking step.

Fineblanking involves three press motions, which distinguish fineblanking presses from conventional blanking presses. It uses a triple action type of press which is very rigid in construction and fully hydraulic. The parallelism of the upper bolster and the ram table is maintained very closely since there is practically no clearance between the punch and die, and an out-of-parallel situation would quickly result in a shortened tool life.

Formerly, all fineblanking tools were compound, now a growing percentage are progressive. That means some tools will have off-center loads, so the press must be rigid. Two features of the fineblank press stand out. First is the adjustable solid stop for setting the top dead center or shut height of the tool. It is important to keep the punch from entering the die. If the punch is continuously extended into the die opening, the punch or die will eventually crack. The solid stop can be set to ensure the repeatability of the shut height. The second feature is the structural support the machine gives the tool. If the tool is allowed to "breath" under pressure, tool life will be shortened. Stroking rates vary from 35–120 per minute.

MACHINE FEATURES

In the family of stamping processes, which include blanking, punching, coining, drawing, bending, etc., there is a variety of different machines to handle the different applications. The fineblanking process also requires its own special machine.

The following features are generally available.

A tool safety to protect against scrap or parts not properly ejected.

Hydraulic tool clamping for quick tool changing.

CNC control to program closing speed, shearing speed, counterpressure on the ejector punch, opening speed, top tool height, and V-ring force, and at the same time to keep statistical information regarding the tool, the press, and the operator. CNC also controls the compressed air jets for workpiece and slug removal and operation of the stock feeding device.

To go along with CNC, provisions for quick die changes have been made. This further reduces downtime and makes for a more flexible system.

An important development was the introduction of coatings on tools to resist wear. The coating (cf. Chapter 4) extends the useful life of the tool because it creates a surface hardness of about 70–80 Rc and adds lubricity to the surface. Lubricants have been developed especially for fineblanking (it is always wise to use lubricants which have been prepared for a specific task; this is true whether the process is turning, stamping, forging, or drilling).

The hydraulic fineblanking machines come in sizes from about 200 to 2500 tons, while the few mechanical machines which are available range from about 25 to 250 tons. Many companies are looking for vendors who can supply all good parts so they can reduce inspection costs. Fineblanking fulfills this requirement because it is by nature a technique oriented to quality—quality is built into the process right from the start.

Fineblanking routinely maintains accuracies of 0.001″ on all dimensions: hole sizes, center distances, form size, and all locations. Generally, there is no difference between the first piece struck and the one-millionth. Surfaces sheared will probably have finishes of 32 microinches or better, and certainly no worse than 63.

The major benefit of this process is that a stroke of the press can deliver a finished part. Normally, all that is ever required as the parts come off the press is a vibratory tumbling or high speed belt sanding. This eliminates the need for secondary operations such as flattening, milling, counterboring, countersinking, edge grinding, and reaming.

There are other benefits to the process. Small holes, which are generally drilled after a stamping operation, can now get put in during fineblanking. In conventional stamping, the smallest hole must be at least $1\frac{1}{2}$ times the material thickness. The same minimum restriction applies to spacing between holes and closeness to edges of holes. In fineblanking mild steel, the figure drops to 60% or 50%. In copper or aluminum, the figure drops to 35%. The same holds true for narrow slots.

Sometimes parts which ordinarily would be either machined from the solid, or conventionally blanked and then machined, can be fineblanked completely in one step.

All presses have tool safety devices built in to prevent cycling to the next stroke until all slugs and parts have been cleared from the die. That means two things: low tool breakage and less expensive tools because the tools do not require separate safety devices.

If greater strength is required for a part processed in a fine-blank machine, all you have to do is increase the part thickness. The fineblank tool requires no change. This wouldn't be true if the part was being stamped in a conventional machine. Semi-piercing, which is a partial extrusion of holes, can be fine-blanked along with the rest of the design. There are times when this is an important function. At the same time, you can counter-sink, counterbore, chamfer, and coin; and, of course, you never have to flatten the part.

On the down side, there are a few things to consider. When working with tool steels, alloyed steels, or exotic metals, all limits previously stated must change. Tolerances and minimum hole sizes increase slightly. Tests will indicate what to expect from a specific material; and sides are not perfectly square as stated. Up to a 1° taper may have to be added; that means a piece of metal ¼" thick could have as much as 0.004" taper. Die roll, which is a slight rounding of edges that occurs on all stamped parts, is also experienced. It happens on the side which enters the die opening; so, if you must have a sharp edge, make it the punch side of the part.

ADDED VALUE

Fineblanking permits designers to incorporate many additional part features at no extra cost. In the same stroke which creates a part, the process can make inside diameter marks, coined sections, self-rivets, contact points, bosses, spacers, female cam tracks, assembly pins, countersinks, and a variety of other functional shapes which would otherwise require machining.

Single station fineblanking tools produce parts with forms, bends, and offsets. These parts are consistently within specifications. Since the part features are created as the tool closes, the dimensional accuracy is not affected and the features are not distorted.

A wide variety of metals with good cold-forming characteristics lend themselves to the process. These include carbon and alloy steels, stainless steels, brass, aluminum, copper, bronze, and monel. Other materials can also be processed depending on thickness and configuration.

Components traditionally considered too heavy for conventional stamping (in excess of ½″ thick) can be fineblanked at a definite cost saving over machining. Even the heaviest fineblanked parts have perpendicular, sheared edges and straight

Fig. 12.1. Examples of parts produced by fineblanking.

walled holes with superior finishes. When part function requires full bearing sidewalls, expensive milling, reaming, or grinding operations are not necessary.

The unique combination of triple action fineblanking presses and "zero clearance" tooling produces thousands of parts with repeatable accuracy and uniformity. Tolerances of 0.001"– 0.002" for size and position are commonplace, and 0.001" per inch for flatness can be held when required.

Designers can enjoy opportunities beyond the scope of conventional stamping. For instance, they can fineblank holes as small as 40% of material thickness, and interior openings can be located more closely to each other or to edges without secondary operations. Designs with intricate or fragile cross sections can be reproduced with maximum integrity and consistency. (See Fig. 12.1.)

If parts made by another process are not performing satisfactorily, or are too expensive, consider a change to fineblanking. Remember, however, that it takes time to learn about a new process. Rather than take a chance of overlooking possible benefits, it would be a good idea to call on the experts for advice.

13

COLD FORGING — COINING

PROCESS

Cold forging is a process which forms metal at room temperature to a desired shape by forcing a lubricated slug —into a closed die under very high pressure. The force could be generated by hydraulic or mechanical presses. The process deforms the metal plastically. In this process, the metal is reformed in continuous, unbroken lines that follow the contour of the part, increasing resistance to shear and breaking.

In cold forging, the slug (blank) contained in the die is shaped by compressive force of a punch. The metal may be forced back along the punch as a backward extrusion, or it may be forced forward through the die as a forward extrusion, or it could be a combination of both.

Cold extrusion is a high production operation which can be automated to produce complete or near complete parts without generating waste material. These cold extruded parts begin as raw material in rods, bars, coils, or billets. The individual pieces, called slugs, are produced by shearing, sawing, blanking, or any other convenient method.

Since blanks are placed into closed dies, the blank volume is important. Automatic sorters can be used to separate blanks by their gram weight into over, under, and correct weight. A slight adjustment of the press can compensate for slight changes in weight, as long as all slugs in each run are the same weight.

The part may, in some cases, be used just as it comes from the press, or further secondary processing such as machining or annealing may be required. Sometimes the blanking step may work harden the piece. It will then require annealing before further forming can take place. Sometimes two forming steps are necessary with an intermediate annealing step.

LUBRICATION

There is a considerable force which causes heat, and resistance to movement, when pressure is applied to the slug in a die, which is why the slug must be lubricated. The inside surface of the die must be protected. The protection is provided by dipping the slugs in a solution of zinc phosphate. It is important to give the slug a definite amount of lubricant: too much would take room away from the slug and it wouldn't flow properly to fill the die; too little would prevent a free flow, and damage to the die might result. Zinc phosphate is far superior to oil because it is dry and nonstaining. It doesn't soil hands or equipment by picking up dirt as oil does, and it prevents rust. Sometimes a part requires another press operation, and it often gets a second zinc phosphate dip.

MATERIAL

Formerly, only low carbon steels were cold forged. Now, almost any ferrous material can be cold forged whether it is stainless, an alloy, or a high carbon steel. Most nonferrous metals can also be cold forged.

Some aluminums are too brittle and they break before forming. Those successfully processed are 1100, 2014, 3003, 6061, 6351, and 7075. Copper cold forges easily with very low tonnage required. Copper parts are often cold forged to finished dimensions without requiring secondary operations.

A cold forged part in low alloy steel can be stronger than a part machined from a high alloy. The ability to substitute low alloy steel for high can save considerable material costs. Automobile companies used to make parts from only familiar materials; now they investigate any possible material change since they use so much of it. Many times, changing the manufacturing process to cold forging eliminates the need for heat treating.

GENERAL INFORMATION

Cold forged parts pressed in moderate-sized machines can weigh as little as a few ounces to as much as 60 pounds. There are a few gigantic-sized presses around that can handle larger pieces. Parts can be round, flat, long, short, or almost any shape. Various wall thicknesses can be accommodated;

nevertheless, vendors prefer that the bottom of a cylinder be thicker than the walls, as is the situation with projectile cases.

Grooves and undercuts are normally machined in after the forging is pressed. Concentric parts are commonly pressed, but eccentric parts can be made as well. Critical dimensions can be held in this process. The accuracy is machined into the die which can hold the accuracy for hundreds of thousands of parts.

The weight of slugs can be held to ± 0.5 grams. Flatness can be held to 0.005" across a 5" diameter surface. Tolerances of ± 0.003" can be held on diameters, and wall thicknesses of parts up to 2" in diameter. Center distances between holes can be held consistently close.

Repeatability of cold forged parts is constant, whereas in machining, the dulling tool will vary dimensions. The surface finish of cold forged parts can be held to 30 microinches; in copper it can be held to 10 microinches.

Depending on part complexity, lead time can vary from 4 to 24 weeks. Normal lead time is about 16 weeks. This, of course, is time required to make dies. Since a die to make 100 parts could cost 50–100% as much as a die to make 1,000,000 parts, this process should not be considered for short runs. The initial tool cost is the customer's; after that, maintenance is the vendor's responsibility.

Physical Properties of Steels Before and After Cold Forging

	Before Forging			After Forging	
AISI Steel	Rc B	Tensile Strength (psi)	Yield Strength (psi)	Rc	Tensile Strength (psi)	Yield Strength (psi)
1010	58	52500	35950	88	91250	85300
1020	57	58000	40250	92	104800	101166
1040	81	74700	49750	94	110150	97875

COLD FORGING VERSUS COINING

Just about everything said about cold forging is true about coining. In addition, coining is the forming of metal surfaces by a compression force sinking or raising the product material to the relief features of a tool and die as in making coins (money).

The work table of a coining press is smaller than that of a forging press. This allows less deflection of the ram support, sometimes called "breathing." With less "breathing," smaller dies can be used and greater accuracy can be expected. This is one reason why you are urged to consult with vendors. If they can figure a way to make your part by coining or cold forging, there generally is a cost saving for you.

Recently, an innovative coining vendor, Cousino, of Toledo, OH, deliberated the possibility of making a copper part shaped like a nose cone. The inside and outside walls had to be concentric to within 0.002", and the tapered diameters required were mathematical equations (polynomials) whose dimensions had a total tolerance of 0.002" each. Cousino was able to complete this job successfully by using two sets of tools for two separate operations. This process was cost effective when compared to hydroforming plus machining, or orbital forging plus machining. Very often, the bottom line (cost) is a reflection of the vendor's ingenuity.

14

HYDROFORMING

PROCESS DESCRIPTION

This process is sometimes referred to as fluid forming or rubber diaphragm forming because of the rubber, fluid pressure forming chamber which acts as the upper die member. The process was developed just after World War II as a method of making small quantities of deep drawn parts at less expense than the conventional method of deep drawing in a hydraulic or mechanical press.

Hydroforming uses a unique forming process which offers many advantages. The fluid-filled forming chamber serves as the upper blankholder and female die element — a universal die which accommodates any shape. (Normally a thick, neoprene pad is cemented to the diaphragm to protect it). In the lower portion of the press is a punch attached to a hydraulic piston, as well as a blankholder ring (draw ring) through which the punch moves.

At the start of the machine cycle, the top of the punch is level with the top of the draw ring. See Fig. 14.1A. The blank to be formed is on the blankholder ring and the punch just touches it. The forming chamber is lowered until the wear pad hits the blank, at which time a small hydraulic force is applied to the blank. This force may be as high as 5000 psi. See Fig. 14.1B.

The punch moves upward under increasing pressure (up to 15,000 psi). It pushes the blank through the flexible die member forcing the blank to assume the shape of the punch; the diaphragm actually wraps the blank around the punch. All during the cycle, the blank is held securely by the two opposing forces. See Fig. 14.1C.

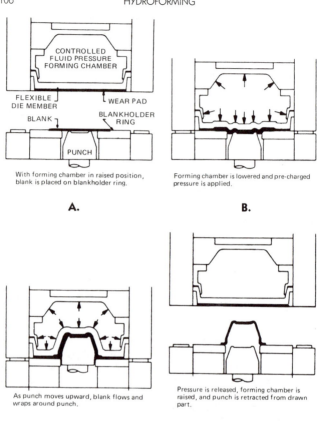

With forming chamber in raised position, blank is placed on blankholder ring.

A.

Forming chamber is lowered and pre-charged pressure is applied.

B.

As punch moves upward, blank flows and wraps around punch.

C.

Pressure is released, forming chamber is raised, and punch is retracted from drawn part.

D.

Fig. 14.1. Hydroform press operating cycle sequence.

To complete the cycle, all pressure is released. The forming chamber rises and the punch is retracted from the completed part. See Fig. 14.1D. A Dayton Rogers or similar trim tool may be used to remove unwanted material such as the flange. (Normally, the radius adjacent to the flange is formed to be large enough to facilitate this cutoff.)

COMPARISON TO CONVENTIONAL FORMING

In a conventional press, the blank is positioned on top of the lower die, and then the draw ring is lowered to contact the blank. The upper die is then lowered to contact the blank and the part is drawn. Throughout most of the cycle, the blank is not touching one die. Sometimes this leads to tears or wrinkles because the blank is "out of control." At the same time, the part is stretching and work hardening.

Often, more than one draw is required in conventional deep drawing with annealing steps in between. In hydroforming, most asymmetrical parts can be made in one stroke without annealing. Since only a draw ring and a male punch are required for the hydroform, 40–50% of the usual deep draw die cost is saved. The hydroform setups are simple and fast because there is no die set. The tooling is self-centering and self-aligning. Practically all sheet metals capable of being cold formed can be hydroformed. This includes carbon and stainless steels, aluminum, copper, brass, precious metals, and high strength alloys.

When using a hydroform, hydraulic pressure in the forming chamber keeps the diaphragm in contact with one side of the blank while the punch, of course, is in contact with the other side of the blank. This keeps the blank in constant control and allows a smooth flow without scuffing, making stretch lines, or work hardening. After an R&D program, in which a part is perfected by the spinning process, the spinning chuck can be used as the hydroform punch, and parts can be made by this faster process. Since many draw rings are stored for ready use, you generally don't have to pay for one of them.

Irregularly contoured parts are easily formed using the hydroform process because matching dies are not required. Since hydroforming flows rather than stretches the metal, there is minimal material thinout. Sometimes this results in a worthwhile cost savings in material, especially when expensive alloys are used. There is a considerable saving in material for tools. Instead of expensive tool steels, this process uses cast iron or plain machinery steel even for long runs. As a matter of fact, kirksite or cast plastics can be used for very short runs. And finally, if the thickness of the material is changed, no tooling modifications are needed. (See Fig. 14.2.)

Fig. 14.2. Typical parts produced by hydroforming. (Courtesy of Roland Teiner Company, Inc.)

PROCESS EVALUATION

Although hydroforming enjoys distinct advantages over conventional deep drawing, certain basic factors should be evaluated before selecting it. The part should be analyzed to determine which factors create the most cost. Normally, for an order of 2000 parts or less, tooling will be the largest cost component. For an order of 20,000 parts, tooling and material costs will be about equal, with direct labor now creating a significant portion of the cost.

As quantities increase, the labor cost becomes more important. Accordingly, total manufacturing costs should be investigated before the process selection is made. That means you should evaluate the following expenses: material, tooling, die design, manufacture, development and setup, all labor, direct and indirect, and any finishing costs required. For example, a spinning might take 15–30 minutes to produce, whereas the same part may require only 1 minute to hydroform. Generally, hydroforming can fashion a part without annealing it, but spinning the same part might require 1–3 anneals. Many odd-shaped parts can be hydroformed, but only symmetrical parts can be spun.

For price comparison purposes only, the following generalizations are offered. On the average, spinning tools cost about one-tenth as much as draw press tools; hydroforming tools cost from two to four times as much as spinning tools, including the small additional cost (perhaps $200) of a draw ring. If a draw ring is available from stock, sometimes spinning and hydroforming tools cost about the same.

EQUIPMENT TERMINOLOGY

Hydroform presses are referred to by the size of the flat, circular blank to be formed, the maximum forming pressure, and the depth of draw. If a press can handle up to 8″ in blank size, it is designated as an 8″ machine. Draw depths can go up to 12″. Forming pressures vary from 5000 to 15,000 psi. If a machine can handle a 12″ diameter blank, can use a forming pressure of 15,000 psi, and can draw a blank 7″ deep, it is referred to as a 12-15-7 hydroform press.

HYDROFORMING DESIGN SUGGESTIONS

Try to design to the inside of the part because the part wraps itself around the punch. If you disregard metal springback, the inside of the part should have the same dimensions as the punch. If the metal is seen to spring back, the springback amount of stock can be removed from the punch in steps until the required dimensions are obtained consistently.

Remember that the open (flanged) end of the part will remain as thick as the starting blank. This may be important to the next step whether it is welding, assembly, or whatever.

Zero draft can be used if necessary. However, it is prudent to use a draft of 1 or 2 degrees to ensure longer tool life (less scratching and galling) and allow faster machine cycles.

In hydroforming rectangular pieces, the print might call for tighter corners than you get. At all times, the minimum radius should be 2–3 times stock thickness. However, this radius can be decreased with a little extra work. Just lower the punch about $1/16''$ and push the diaphragm down swiftly; this will produce a sharper corner.

An unusual advantage of hydroforming is in the simplicity with which certain drill jigs can be made. If a part requires drilled holes, it can be hydroformed first. A heavier piece of material is then hydroformed directly over the finished part. The finished part is then removed and the larger part has the holes laid out and drilled. Thereafter, the larger part is used as a jig to locate holes in the finished parts.

Tight tolerances cost money whatever the manufacturing process used, so common sense dictates loose tolerances where possible. Draw depth tolerances are typically $\pm 0.020''$, but ± 0.010 can be held. Inside diameters of $\pm 0.005''$ or even $\pm 0.002''$ can be obtained. Often it is the material that determines what tolerance to request.

Hydroform vendors should discuss the tools they are designing for your job. Although they must see that those tools deliver the parts as required by the drawing, there is much knowledge you can acquire from these discussions.

15

METAL SPINNING

HISTORY

History records the fact that the Egyptians practiced an archaic form of metal spinning, which makes it the oldest known method of producing hollow, circular metal components. It was introduced into our country in the middle of the nineteenth century. At first, the process was used solely to produce parts made from soft, easily worked, nonferrous metals. It wasn't until shortly after World War II that harder, tougher, ferrous metals were worked. Now, due to its versatility, metal spinning is used for small- or medium-sized production lots.

PROCESS DESCRIPTION

In this process, a metal disc is spun at controlled speeds on a machine similar to a machine lathe. The disc is held between a mandrel, secured to a chuck (of sorts), and a follower attached to the tailstock. The mandrel corresponds to the inside contour of the part to be produced. Power is used to revolve the mandrel, disc, and tailstock follower.

Spinning tools are then forced against the rotating disc, traditionally by hand. The operator forces the blank to assume the shape of the mandrel by means of a series of strokes. The metal flows in a manner similar to clay on a potter's wheel.

THE OPERATION

The size of the metal blank is determined by the surface area of the finished part. Pressure on the spinning tool is generally exerted by hand, although some machines have air power or hydraulics to perform this task. As a matter of fact, some automatic machines have template controlled devices which speed up the operation somewhat.

Normally, parts for spinning are symmetrical and circular in shape because the machine rotates. The process is so simple and basic that pieces as large as 26 feet in diameter and thicknesses up to 2″ have been worked. But most of the work processed is considerably smaller.

The first step in spinning is to make the mandrel. Nowadays, compressed wood or laminated wood is often used for the mandrel. Sometimes, especially if quantities warrant it, metals could be used. The mandrel is secured to the chuck; in fact, the mandrel itself is sometimes called the chuck. The blank, which is a flat metal disc, is centered and clamped tightly between the mandrel and the tailstock or an extension of the tailstock.

The lathe-like machine turns both mandrel and disc as a bar-like tool is used to flow the metal over the mandrel. Lubricants are generally used to counter the friction of the process. If the disc is thin and large in diameter, it may have its edge curled to establish a higher degree of rigidity.

The operator must possess a reasonable skill in order to produce uniform parts. He exerts pressure on the disc, controlling a long, blunt tool with his hands, arms, and body. He also uses special tools to "finish" the surface, trim off excess metal, apply beads and flanges, and make parts with fairly close tolerances considering the fact that this is all hand work.

A sophisticated variation of this spinning process is called shearforming, and it is described in Chapter 16. Shearforming is different from spinning in that it achieves a deliberate and controlled reduction in disc thickness and thus can make shapes which the conventional process cannot. It also speeds up the spinning process tenfold.

ADVANTAGES OF METAL SPINNING

Due to the mechanical working of the disc in spinning, the grain structure is refined and thus provides metallurgical benefits. The heavy forces required to plastic flow the metal during spinning orient the grains parallel to the principal axis. This phenomenon is similar to the grain orientation caused by forging.

According to many tests, cold working the metal during

Table 15.1 Typical Dimensional Tolerances that Can be Attained with the Spinning Process

	As Spun — For Most Commercial Applications	Special Applications
Up to 24" Diameter	± 1/64" to 1/32"	± .001" to .005"
25" to 36" Diameter	± 1/32" to 3/64"	± .005" to .010"
37" to 48" Diameter	± 3/64" to 1/16"	± .010" to .015"
49" to 72" Diameter	± 1/16" to 3/32"	± .015" to .020"
73" to 96" Diameter	± 3/32" to 1/8"	± .020" to .025"
97" to 120" Diameter	± 1/8" to 5/32"	± .025" to .030"
121" to 210" Diameter	± 5/32" to 3/16"	± .030" to .040"
211" to 260" Diameter	± 3/16" to 5/16"	± .040" to .050"
261" to 312" Diameter	± 5/16" to 1/2"	± .050" to .060"

Courtesy American Metal Stamping Association

spinning increases tensile strength appreciably. Sometimes this increase actually doubles the strength. The spinning process allows the contour to be formed with little or no added machining. A cast part would probably have to be thicker to allow for a machine finish because we cannot cast thin sections. See Table 15.1.

TOOLING

Much of the accuracy of the spinning job depends on the mandrels. Carefully selected woods, properly kiln-dried and properly glued together, make excellent mandrels. Even if metal mandrels are required, they are considered low cost compared to the usual tools for a draw press.

A comparison between two identical parts, one made by spinning and the other by press forming, illustrates the steps saved by spinning. To make the cone illustrated, spinning requires 3 steps while press forming takes 8 steps. See Fig. 15.1. And spinning is an ideal process to select when production quantities do not justify draw press tooling. This is especially true when dealing with large diameter parts.

Consider the three basic shapes: cone, hemisphere, and straight-sided cylinder. (See Fig. 15.2.) The cone is a simple

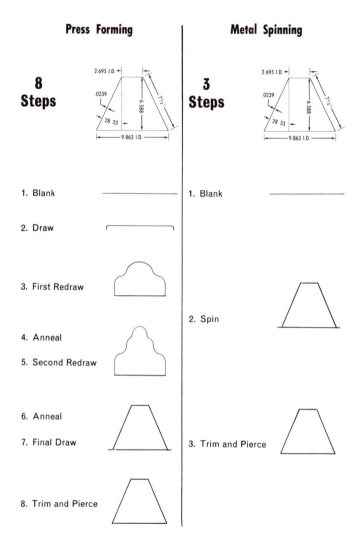

Fig. 15.1. A comparison of the press forming and metal spinning processes. (Courtesy of American Metal Stamping Association.)

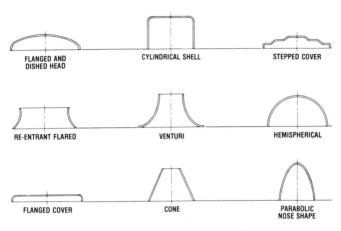

Fig. 15.2. The basic metal-spun shapes. (Courtesy of American Metal Stamping Association.)

shape for spinning, while it is difficult for press forming. The hemisphere is more difficult to spin, but it is still easily done. Spinning a sharp angle exposes the metal to a great strain and requires more skill. Consequently, the straight-sided cylinder is the most difficult of the three shapes to spin and requires a more skilled operator.

Sometimes parts are spun in preparation for a machine finish, and there are times when parts are made by other processes and then finished by spinning. For instance, it is possible to achieve lower cost by rolling sheet metal shapes and seam welding them into cylinders; then the parts could be completed with a spinning operation. There is a multitude of parts, like those shown in Fig. 15.3, which are obvious candidates for spinning.

The message from spinning vendors is short and to the point: if you have a small quantity of symmetrical sheet metal parts to make, consider spinning them. In fact, it may pay to consult with a spinning vendor even for a moderate quantity, especially if you have a tight schedule.

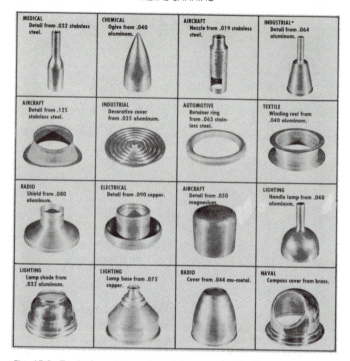

Fig. 15.3. Typical parts produced by metal spinning. (Courtesy of Roland Teiner Company, Inc.)

16

SHEARFORMING

BACKGROUND

A form of metal spinning more precise than that used by the Egyptians originated in China around 900 AD, and it was taken to America about 900 years later. At first, the equipment was crude and the operators had to be craftsmen to produce even the simplest parts. (The preceding chapter described metal spinning.) However, as time passed, the operators wanted to work thicker and stronger materials. This created a demand for mechanical advantage to replace the brute strength of the operator, and this was the beginning of power spinning. It introduced hydraulics to provide greater and greater forces.

A modern variation of the spinning process is known as shearforming (also called flow forming or floturning). This is a sophisticated version of the ancient art of metal spinning. It has the added ability of coping with appreciable variations in wall thickness; and close tolerances are easily obtainable by the operator through the machine controller.

In metal spinning, a flat or almost flat blank of sheet metal is forced by an operator, leaning on a long-handled forming tool, to conform to a convex mandrel. In the process, the wall thickness of the spun part is reasonably constant, except for some stretching which occurs where the blank is bent. This thickness change is not sought after — in fact, it is normally undesired — but it is accepted as an inconvenience of the process.

Shearforming, on the other hand, allows the creation of thicknesses which vary from point to point by as much as 100%. Another constraint of spinning involves blank thickness and radii. In spinning, blank thickness is generally less than $\frac{1}{8}''$, and most workpiece radii are more than 5 times blank thickness. Normally, shearforming machines can handle blanks much thicker, and can bend any radius the material itself can endure. Of course, this depends somewhat on machine size. However, both processes can, under the proper circumstances, form steel plate 1″ thick.

PROCESS

Although the process of spinning has been accelerated by several innovations, it has a few disadvantages which have been overcome by the new shearforming equipment and the technical advances of that science. Before describing the actual process, a review of these advances is in order.

In shearforming, the forming rolls are controlled by strong, hydraulic servo systems. The machine structure is extremely rigid. In fact, if the proper size machine is used, the major limitations in shearforming are the physical properties of the blank, not the machine.

Whereas the earlier shearforming machines required the rollers to travel in a straight path, subsequent models incorporated hydraulic tracer units which allowed shapes other than cylinders or cones to be formed. And the newest models create parabolic or curvilinear shapes provided that the initial blank is designed on the basis of the sine law (which will be described shortly).

The word "blank" is applied to any part or form of metal secured to the chuck, preparatory to shearforming. Normally, we think of a blank as being a piece of flat sheet metal, either round, square, or some other shape, depending on the finished contour required. But it can be a cupped blank or cylinder produced by spinning, pressing, forging, or casting. It could also be a part which started as plate, was rolled, welded, and rough machined before being shearformed. A very popular product for flow turning (also called floturning, as in the Lodge & Shipley machine, Floturn) is a conical shape. But there are two limiting factors when considering conical work.

The first limiting factor is the amount of reduction that the material selected will take without requiring annealing. Plain carbon steel, some steel alloys, annealed aluminum, and some stainless steels (like the 300 and 400 series) will endure a 75% reduction. As will be discussed further along in this chapter, the amount of reduction is strictly governed by the sine function of the side angle of the cone. The sine of 30 degrees is 0.5000 and the sine of 15 degrees is 0.2588. So, when starting from a flat blank, if the cone's side angle is 30 degrees (60 degrees included), the thickness reduction would be 50%. If the side angle were 15 degrees, a 75% reduction will occur. Some of the high strength alloys cannot withstand such a large reduction. In fact, some materials require heat when being floturned.

The second limiting factor is the capability of the size of the machine to form the material selected. This limitation dictates the maximum blank thickness that a specific machine can form, as well as a maximum reduction. For example, even if a material can withstand a 75% reduction, if a blank too thick for a machine is positioned on it, the machine is not powerful enough to do a 75% reduction in one pass. The machine could probably perform the task in two passes.

When shearforming cylindrical work from a pressed blank, it is sometimes necessary to machine the ID and OD prior to shearforming; otherwise, the process might result in a varying wall thickness. (See Fig. 16.1.)

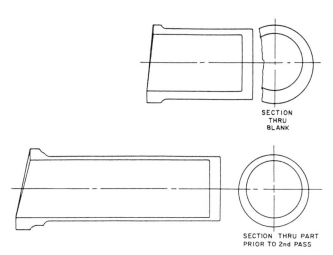

Fig. 16.1. Prior to shearforming cylindrical work from a pressed blank, it is sometimes necessary to machine the inside and outside diameters of the blank.

Once the procedure for shearforming a part is established, the process is quite simple. For any subsequent run, the setup is as follows.

1. Mount the mandrel on the spindle faceplate or chuck and true it up.
2. Mount the selected (specific for the job) roller on its spindle.
3. Adjust the slide to the correct angle relative to the mandrel, and position the roller at the correct distance from the mandrel.
4. Set the microprocessor for the calculated rate of speed and feed and all the travel limits.
5. Finally, turn the selector switch for the desired cycle: manual, semiautomatic, or fully automatic.

When those steps have been taken, a typical working cycle would be as follows.

1. Load the blank into the machine. Sometimes a centering device is used to assist this step.
2. Use rapid advance to bring the tailstock to the clamp position.
3. Depress the operating button which turns the mandrel, supplies coolant, and begins cross slide movement to the mandrel.
4. The cross slide's saddle feeds automatically for a predetermined distance to form the part.
5. The cross slide retracts rapidly to the start position.
6. The spindle stops turning and coolant/lubricant stops flowing.
7. The tailstock goes into rapid retraction and the finished part is pushed off the mandrel by an ejector rod which protrudes through the mandrel.

Results from shearforming will vary according to the work hardening characteristics of the material and the magnitude of the residual stresses in the finished part. The quality of the finish will vary also, depending on the combination of blank size and machine feed and speed. The materials will react in a manner similar to that obtained in deep drawing.

MACHINES

A typical shearforming machine looks much like a lathe. Its motor-driven heavy-duty, precision spindle holds a mandrel contoured to fay with the internal configuration of the part to be shearformed. The workpiece blank, which is generally flat, is held against the front face of the mandrel by a small-diameter rotating push rod, extending from a hydraulically actuated tailstock.

In Sweden, the Hallarydsverken Company built the first spinning-type machine that does what we call shearforming. Then, in Germany, the Leifeld, Bohner and Kohle, and the Keilinghaus Companies produced machines for sale. The Swiss Jenney Pressen Company and the American firms of Lodge & Shipley, Cincinnati Milacron, and Autospin also produce machines. Machines are also made in England and Japan.

All major parts of the machines are generally made from a good grade of cast iron. An important characteristic of cast iron is that it absorbs vibrations; that is the reason machine builders use the material so generously. These machines have several design features in common: the headstocks always have large roller bearings to withstand the high axial and radial loads imposed; hydraulic power is generally used to operate the tailstocks and cross slides; and the machine's main parts are very robust to withstand severe vibrations.

Although many products can be made in a single slide machine, most machines now seem to be multiple slide types. Twin slides are necessary to avoid axial deflections. Three- and four-slide machines are also made to improve production of specific products.

The transmissions may have high- and low-speed ranges going from perhaps 10 to 3000 RPM, depending on the size of the machines and their duty roll. Since there is sometimes difficulty removing parts from the mandrel, there is an ejector rod, actuated hydraulically, in the hollow spindle to push parts off. The stopping distance of the roller from the chuck has been consistently maintained by a micrometer dead stop which was operated by a handwheel. This assignment can now be performed via the microprocessor.

The process generates considerable heat so that adequate cooling is of major importance. A significant responsibility is the selection and maintenance of the fluid which not only cools but lubricates. One highly successful shearforming vendor insists that his treatment of the fluid used for this purpose is instrumental in his success. Naturally, his treatment is proprietary and certainly contains more than filtration.

The earlier shearforming machines derived much of their functional advantage from the use of templates. Once the optimum operating adjustments were made, the movements were consistent and production was good. Now, of course, the microprocessor is set and reset until the ideal conditions are present, precluding the necessity of templates.

When highly accurate parts are needed, care must be taken to use a mandrel turning absolutely concentrically with the chuck. Sometimes this is ensured by finish grinding the mandrel on the machine with a portable grinding unit.

The machine manufacturers realized that too much floor space was required for machine size when they contemplated the very large parts that were being designed. In order to work the large diameters and larger lengths, it was decided to change from horizontal machines to vertical, just as the industry had done many years ago with the lathe. The builders retained their insistence on extreme rigidity and strength; this is absolutely required when one considers the fact that as much as 250,000 pounds of force can be exerted by a machine which may be powered by a 350 HP motor. See Fig. 16.2.

The earlier shearforming machines depended on microswitches, trip dogs, and solenoid valves to control the movements and operations of the roller slides, carriage, tailstock, and templates. These devices regulated the flow of hydraulic fluid to the rams and cylinders which caused the movements. The newer machines use microprocessors to control these movements.

FORMULAS

The cone is the basic shape on which the shearforming process was founded; thus, a formula based on the sine law is derived from the cone. The sine law presents a relationship between the thickness of the blank from which the part is to be formed, the included angle of the cone, and the wall thickness of the finished part.

The formula can be expressed in two ways: as the sine function of the angle subtended by the cone center line and its wall; or as the cosine function of the complement of the above angle.

Fig. 16.2. A vertical shearforming machine. (Courtesy of Autospin Inc.)

The expressions are written:

$$T = \frac{t}{\text{sine } \propto} \qquad T = \frac{t}{\text{cosine } \varnothing}$$

where

T = thickness of required blank

t = wall thickness of the cone

$\propto$ = angle subtended by center line and cone wall

$\varnothing$ = angle subtended by cone wall and base line extended.

(See Fig. 16.3.) These formulas state that the depth of any conical part made by shearforming is not dependent on the diameter of the blank alone, as would be the case in spinning or deep drawing, but on the blank thickness also. The diameter of the blank is, in turn, dictated by the major diameter of the finished cone.

As the sine angle of the cone decreases, the thickness of the required blank increases. When the sine angle is 15 degrees,

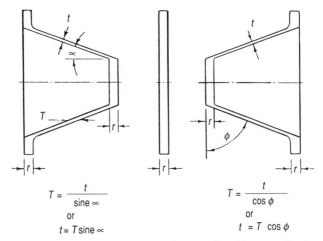

$$T = \frac{t}{\text{sine } \propto}$$
or
$$t = T \text{ sine } \propto$$

$$T = \frac{t}{\cos \phi}$$
or
$$t = T \cos \phi$$

Fig. 16.3. Basic formula for the sine law (angle with center line) on left, and cosine law (angle with base line extended) on right.

making the total cone angle 30 degrees, a second shearforming pass should be planned. A different formula is used to calculate wall thickness when the included angle of a cone is less than 30 degrees. This formula also is based on the sine law. The cone angle and the thickness of the blank for any wall thickness are interdependent, and altering one will vary the other. Whenever it becomes necessary to make two or more passes to form a cone, it is necessary to use a different mandrel for each pass. (See Fig. 16.4.)

Fig. 16.5 shows how a very small-angled straight-sided cone could be formed from a drawn blank. Two floturn operations were used so that the depth of the blank could be minimized. The part shown in Fig. 16.5 was made from 6061 aluminum, and the blank was in the annealed condition. After the first pass, the material was solution treated (T4) and then the second pass was made. This method of processing takes advantage of the delayed hardening characteristic of the material (age hardening); and it avoids problems dealing with heat treatment deformation. Final operations involve trimming both ends of the part.

Fig. 16.6A, B, and C is used by the Floturn manufacturer, Lodge & Shipley, to illustrate the function of their machine. The sketches also facilitate understanding of the sine law.

To comprehend conical part machining, you must understand the sine law. Fig. 16.6A illustrates the basic elements of machining a cone. Fig. 16.6B explains the axial displacement of material and shows that the original blank thickness is transposed parallel to the centerline, and any increment of material which is at some distance x from the centerline remains at the same distance x in the completed cone. The normal thickness of the cone is reduced by a percentage as determined by the sine law. Fig. 16.6C illustrates how a blank is floturned into a cylinder.

The sine law cannot be used for forming cylindrical parts; those parts could be formed accurately and satisfactorily by first forming a shallow cup and then shearforming in the usual manner. At this time, the thickness of the blank will depend on three things:

1. the thickness of the final wall required;
2. the thickness of the base required; and
3. the length of the finished cylinder.

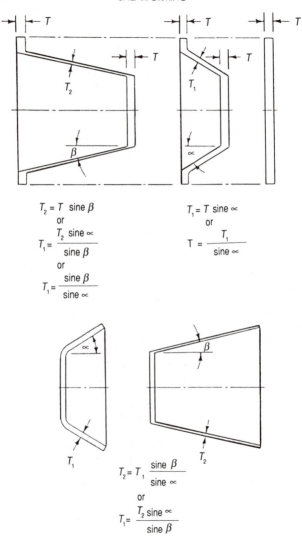

$$T_2 = T \text{ sine } \beta$$
or
$$T_1 = \frac{T_2 \text{ sine } \propto}{\text{sine } \beta}$$
or
$$T_1 = \frac{\text{sine } \beta}{\text{sine } \propto}$$

$$T_1 = T \text{ sine } \propto$$
or
$$T = \frac{T_1}{\text{sine } \propto}$$

$$T_2 = T_1 \frac{\text{sine } \beta}{\text{sine } \propto}$$
or
$$T_1 = \frac{T_2 \text{ sine } \propto}{\text{sine } \beta}$$

Fig. 16.4. The formula for second stage flowing (on left) and conical flowing from a pressed preform (on right).

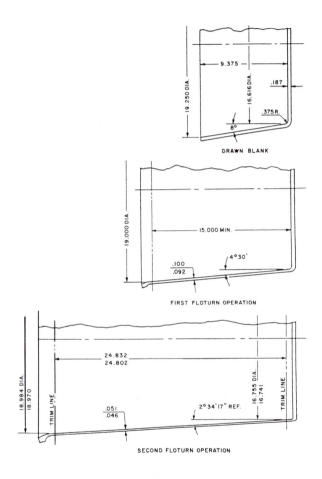

Fig. 16.5. Making small-angled straight-sided cone from drawn blank.

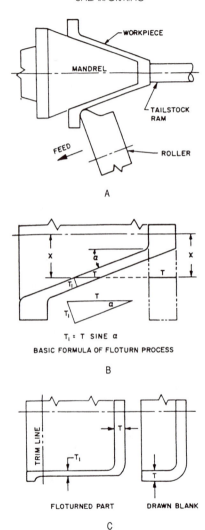

Fig. 16.6. The basic elements of machining a cone. (Courtesy of
Lodge & Shipley.)

The formula generally used for deciding what size blank cup to use is:

$$L = \frac{L_1 \times t_1}{t} \times \left(\frac{dt_1}{dt} \right)$$

where

L = required length of cup or cylinder
L_1 = length of finished cylinder
d = inside of both blank and finished cylinder
t = thickness of blank cylinder wall
t_1 = thickness of finished cylinder wall

The element of the formula shown in brackets is a factor of correction, and it can safely be ignored when considering large diameter cylinders with very thin walls.

When a drawn blank is to be supplied to the shearforming machine and then formed into a contoured part, a computation of the curve blank must be completed first. The blank curve must be such that the relationship

$$\text{sine } \beta = \frac{T_2 \text{ sine } \propto}{T_1}$$

holds true at any point on the blank, where

T_1 = thickness of wall at first flow
T_2 = thickness of wall of final product
$\propto$ = angle of first formed cup blank to center line
β = angle of final product to center line.

A suitable blank curve can be established by treating the part as a series of conical sections and computing the corresponding conical sections of the blank. The computations must begin at the small end of the part. In other words, an incremental solution to the problem is performed.

MATERIALS

Many different materials have been processed by the shearforming technique. Some of these materials, which have been completed in one pass of the roller, would have required a

Table 16.1. Percentage of Reduction for Shear or Tube Spinning Without Intermediate Anneals[a]

| Material | Shear Spinning | | Tube Spinning |
	Cone	Hemisphere	
4130	75	50	75
6434	70	50	75
4340	65	50	75
D6AC	70	50	75
Rene 41	40	35	60
A286	70	55	70
Waspaloy	40	35	60
18% Ni Steel	65	50	75
321 S.S.	75	50	75
17-7 PH S.S.	65	45	65
347 S.S.	75	50	75
410 S.S.	60	50	65
H 11 Tool Steel	50	35	60
6-4 Titanium[b]	55	—	75
B120VCA TI.[b]	30	—	30
6-6-4 TI.[b]	50	—	70
Commercially Pure TI.[b]	45	—	65
Molybdenum[b]	60	45	60
Pure Beryllium[b]	35	—	—
Tungsten[b]	45	—	—
2014 AL.	50	40	70
2024 AL.	50	—	70
5256 AL.	50	35	75
5086 AL.	65	50	60
6061 AL.	75	50	75
7075 AL.	65	50	75

[a]The percentages of reduction, exceeded under specific conditions, are intended as a guide which, if followed, will reduce development time.

[b]Spun hot.

number of cycles in a power press, as well as fairly expensive tooling.

Several aluminum alloys and types of copper perform quite well. Normally, these are processed in the annealed condition. Another material commonly used is Fortiweld, a boronized steel with a tensile strength of 40 tons psi (hence its name). Since this strength is obtainable in the as-welded condition, it is a particularly useful material when weldments are to be shear-formed. Several grades of stainless steel are easily worked. Many high strength alloys have been shearformed although they are difficult to machine by conventional means. Table 16.1 is a compilation of materials and the amount of reduction possible in one pass. (This is only a partial list of materials capable of being shearformed.)

ROLLERS

A massive, hydraulically actuated cross slide supports a roller which could be any suitable diameter from 4″ to 12″. The roller, which is vertical, turning about its center, is often tipped back up to 15 degrees from the spindle axis so as to clear the tailstock push rod.

The leading edge of the roller is radiused in a very specific manner so as to gently create and move a wave of metal on the piece being shearformed. In addition, this gentle metal move-ment is an act of burnishing — it improves the finish on the piece. For satisfactory metal flowing, there must be no tendency for the fibers to be subjected to either a shearing force or a stress which would lift the workpiece from intimate contact with the mandrel. The importance of the roller edge radius is shown in Fig. 16.7A.

If the roller edge radius coincides with the thickness of the metal blank (or is smaller), as in Fig. 16.7B, the radial force tends to hold the workpiece in contact with the mandrel. But as the contact point is followed around the radius in an axial direc-tion, there is no such holding force. In fact, at this spot, there is the beginning of a shearing force. As the shearforming pro-ceeds, the magnitude of the shear force increases and can spoil the finish as well as leave the part in a stressed condition. Fig. 16.7A shows a properly radiused roller edge. Experience indi-cates that the minimum roller edge radius should not be less than 1.5 times the reduction in metal thickness; and, for a more satisfactory result, the radius should be 2–3 times the reduction.

Builders of shearforming machines generally supply a stan-dard set of rollers with each machine. They may be as large as 10–12″ in diameter with large radiuses for heavy duty work, and

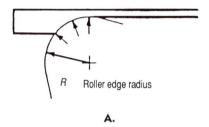

Reactive force
holds metal on mandrel

R   Roller edge radius

A.

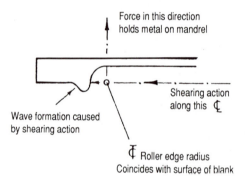

Force in this direction
holds metal on mandrel

Shearing action
along this ₵

Wave formation caused
by shearing action

₵ Roller edge radius
Coincides with surface of blank

B.

Fig. 16.7. Roller edge radius relative to the wave formation.

much smaller for light work. The light duty rollers might have deflector rings on their front face, which bear a large portion of the forming load. As a general guide, a larger radius provides a better surface finish.

Normally, rollers are made from high alloy steel containing molybdenum, vanadium, tungsten, and chromium. This type of alloy can be hardened to a higher Rockwell (65–70) than the mandrel, and is excellent when encountering heat during the manufacturing process.

In any metal spinning operation, the form tool is brought to the center of the rotating blank, close to the tailstock, and works outward toward the operator, forcing the blank against the mandrel. In shearforming, the process is really the same. The roller is brought up to the rotating blank, close to the tailstock push rod; it is then moved out from the center of the blank in a stiff servo-controlled path.

The roller is guided by a microprocessor to form the part whose shape may even be in the form of a mathematical equation. The inner contour is determined by the mandrel, and the outer contour, along with the wall thickness, is established by the processor-controlled roller.

Shearforming, in contrast to conventional metal spinning, is a relatively fast, high-volume production process which shapes blanks with a minimum waste of metal. Once the programming is developed, repeatability of shape and dimensions is assured. Fig. 16.8 shows such an application.

The first shearforming machines produced had only one roller, and yet the work output was tremendous. Since it was computer controlled, the tolerances held by this machine were heretofore unseen in the metal spinning arena. Now these machines are available in either a vertical or horizontal format, and they come with 1, 2, 3, or 4 axes. All are CNC controlled. The more rollers or slides the machine possesses, the more versatile its work output. The second roller was added when it became evident that it was needed to equalize the forming forces of the first roller.

Most of the machines have adapters to permit additional mechanical devices to be attached to perform special functions. There is a device to clamp or center the workpiece; there is one to load and unload parts; and you can get a trimming or beading device. There are also heaters, both gas and induction, for specific assignments, especially for forming extra thick or unusually hard material.

COMPUTER CONTROL SYSTEM

Computer numerical control (CNC) is used in shearforming for the same reasons it is used on numerous types of metal removal machines: it improves accuracy, repeatability, and reduces human error. It also increases productivity, permits the quick setups which allow its economic use even in small lot

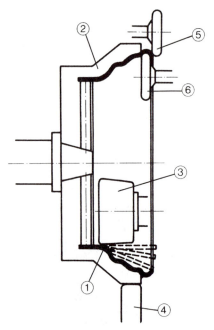

Fig. 16.8. How a wheel rim is produced from a welded steel ring by shearforming. **Forming example:** Forming of a special rim for large earthmoving vehicles from a welded steel ring.

The ring is pushed into the tool and clamped hydraulically. After ① and ② start rotating, the spinning roller ③ forms the steel ring in several programmed spinning passes to the final form. The roller ④ is used to equalize the forming forces exerted on the inside of the tool. After completion of the spinning process, the rollers ⑤ and ⑥ form the flange.

Additional attachments used for the manufacturing of this or similar parts: clamping device, centering device, ejector, and loading and unloading unit.

(Approx. dimensions of spun rim: 60″ dia. Wall thickness up to 1″. Weight of finished rim: approx. 2000 lbs.) (Courtesy of Autospin Inc.)

production, and reduces cost of labor. All the technical knowledge assembled by engineering can be inserted into a database.

The process can be studied in the form of charts and graphic plots even before actual execution. Complex geometries can be formed under CNC control which could not be processed in any other manner (including template control). Can you imagine any other way to control two or four different rollers forming a part simultaneously?

In addition to controlling a gamut of machine functions, CNC monitors several on-line devices such as force or electrical measuring instruments, optical pyrometers, heat sources, and pressure transducers. Papers, which usually accompany parts traveling through a shop, are no longer required because traceability of parts can be an automatic product of a CNC system.

ADDITIONAL ADVANTAGES OF CNC

1. Each shearforming system sold is provided with complete interactive software.
2. The computer executes the software controlling all attached devices, and indicates prompts and messages for the operator.
3. The operator enters numeric data such as machine functions, time delays, feeds, and spindle speeds via a keyboard or numeric keypad.
4. Contours can be optionally entered from a scale drawing via the electronic digitizer tablet.
5. Program data can be printed out on one of several available printers. A multicolored plot can be generated.
6. All program changes and editing can be made at the machine.
7. An unlimited number of spinning passes can be programmed.
8. The system can control any roller or combination of rollers over all necessary spinning contours required to reach the final part configuration.
9. The computer monitors critical machine functions and conditions to maintain optimum safety and mainte-

nance conditions. Machine malfunctions will appear immediately on the screen.

10. The computer-interfaced servo-actuators control position, speed, acceleration, and thrust of the spinning axes to complete the program; yet, the simple design permits years of trouble-free operation.

Apparently, a conclusion which can be drawn from this chapter is that there are many more products out there which could use the shearforming process to economize manufacture.

17

ORBITAL COLD FORGING

PROCESS DESCRIPTION

This process forms a cold metal slug between two dies (upper and lower) with the upper die moving in an orbital motion. This upper die motion creates a high degree of deformation with relatively little force. By means of the orbital motion, the forming force is concentrated in a small area of the upper die. This area is constantly shifted in a certain pattern across the entire piece part surface. The lower die presses the blank against the orbiting upper die. The orbital angle is normally about 2 degrees. This motion results in a progressive forging to final shape, usually in 10–20 cycles and 5–10 seconds. This production method is so promising because accurate piece parts can be formed with a small amount of energy and little loss of material. See Fig. 17.1.

The dies are simple in design and low in cost. For this reason, it is profitable to forge parts in small quantities, sometimes fewer than one-thousand. The flash is usually removed by blanking. Depending on form and application, these parts are finished by chip-making secondary operations.

In the sequence of operations, the slug is placed in the lower die, which is mounted securely in the ram table which travels in an upward motion. The upper die is mounted in the orbital bell in the top half of the press, and during the forming process, it orbits about the orbit angle. The ram then proceeds upward under high speed until it reaches an adjustable preset "Forging Stroke Length." At this point, the ram slows to the adjustable preset forming speed. The upper orbiting die comes in contact with the material and the forming starts. When the press reaches its closed (or top dead center) position, the ram delays for an adjustable dwell time. This permits some finishing orbits

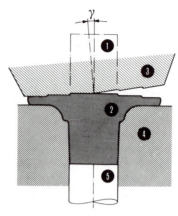

Fig. 17.1. In orbital cold forging, the material (piece part) is forged between the upper and lower dies.

of the upper die to finish "set" the workpiece. The ram then retracts, the ejector is raised to remove the part from the lower die, and the finished part is ejected. The workpiece is thus forged between the upper and lower dies. See Fig. 17.2.

In conventional forging, the pressure is unevenly distributed over the workpiece surface. The pressure increases toward the center of the part and can reach a value greatly in excess of the tool's strength. Sliding friction causes a large part of this force. With an orbiting die in rolling contact against the part surface, there is little frictional force. The metal is allowed to flow radially. This keeps the surface pressure only slightly higher than the material's yield strength.

ADVANTAGES

The orbiting die type of cold forging offers advantages such as short production times, reduced material requirements, increased material strength, smooth finishes, and close tolerances. And if large parts are to be formed, the investment cost would be considerably less than with conventional cold forging.

A 4" diameter steel piece can be forged with an orbital press of 200 tons capacity, whereas a conventional press would re-

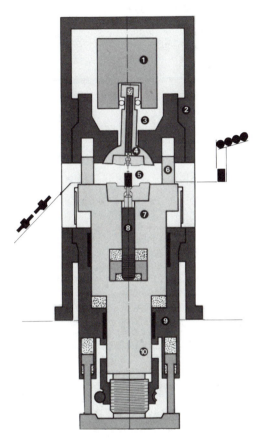

Fig. 17.2. The orbital cold forging press. **1.** Orbital head drive.
2. Press frame. **3.** Upper ejector piston. **4.** Orbital head. **5.** Die.
6. Press ram guides. **7.** Press ram. **8.** Lower ejector piston (inner form
ram). **9.** Cylinder. **10.** Adjustable ram stop. **11.** Piston for ram fast
approach. (Courtesy of Schmid Corporation.)

quire a 2000 ton capacity. Sometimes, thin or complex-shaped parts can be successfully forged with an orbital press, while to do so with a conventional press would be difficult if not impossible. That is because the required compressive load could exceed the strength of the tool.

Cold forming offers the following advantages:

1. elimination of heating equipment, and so less energy is required;
2. the finished part has increased strength;
3. improved accuracy;
4. improved surface finish.

DIMENSIONS AND TOLERANCES

The surface quality of parts is governed mainly by the surface condition of the die. Using polished dies, this process can produce finishes down to 8 microinches. Very often, repeatability and basic accuracy itself is influenced to a degree by deflection in the die. Consequently, the ability to cold form a part with one-tenth the normal force is a significant and worthwhile improvement. See Fig. 17.3 for an example of cold forged parts.

In 1985, a large defence contract was lost because of the deflection in a huge conventional press. A competitor used a coining press, which, because of its smaller table, deflected much less. Orbital forging, because of its technique of applying force, also causes little deflection. Reproducibility of orbital forging ranges between 0.002–0.004".

The grain flow lines in the workpiece are favorably influenced by the cold forming operation which increases surface strength by at least 50%. It is therefore possible to select less expensive, low alloy steels which, after cold forming, display tensile strength values of high alloy steels.

MATERIALS USED

Materials suitable for orbital forging encompass a wide range of both ferrous and nonferrous kinds. Any type of steel, carbon and alloyed, can be formed. They must, however, be ductile and have a spheroidized structure. This type of annealing makes the microstructure more conducive to severe forming. Nonferrous

Fig. 17.3. Parts produced by orbital cold forging. (Courtesy of Schmid Corporation.)

materials like aluminum and copper and high-nickel alloys are suitable for forming, but they must be in a low temper condition.

SLUG PREPARATION

The slugs for orbital forging can start from many forms, including round or profiled rods, tubing, stamped blanks, forged blanks, or preformed slugs. Each slug must be lubricated. A dry lubricant consisting of a phosphate coating, followed by molyb-

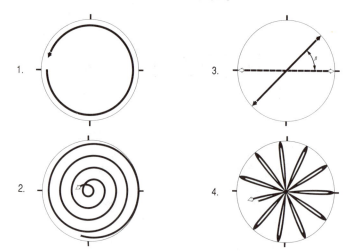

Fig. 17.4. The four orbital motions used in orbital cold forging. **1. Circular motion:** for concentric deformation (the angle of the orbital bell γ is adjustable from 0–2°). **2. Spiral motion:** for radial and axial deformation; well formed centers (γ changes in cycles from 0–2°). **3. Straight pivoting:** for forming in two directions (the angle β of direction is adjustable; γ changes in cycles from 0–2°). **4. Daisy pattern:** especially for forming parts with features on their faces, e.g., bevel gears, clutch parts, etc. (γ changes in cycles from 0–2°).

denum disulphide applied to the slugs before forming, has been successful. This not only keeps the press area cleaner, but also eliminates the possibility of a "hydraulic lock-up" in the tool which can develop when using lubricating oils.

THE ORBITAL PRESS

There are currently three sizes of orbital presses in production: 200, 400, and 630 ton machines. They are fully hydraulic and adjust the shut height of each machine by a mechanical stop. This assures repeatability for dimensional control of the parts being forged.

The ram table and the upper half of the press are guided by four large cylindrical posts. This assures concentricity of the upper and lower dies during forging. The orbital unit consists of the orbital head and gear drive. Within the head rests the orbital bell which holds the upper die. The orbital bell rides in a special spherical bearing designed to absorb the great forces created during forging. The gear drive rests on top of the orbital head. All four orbital motions (see Fig. 17.4) are controlled from within this unit.

Production rates vary from 5–15 parts per minute. Orbital forging represents an important technological advance in the cold forming industry. This particular phase of cold forming produces near net shape more so than any other type of cold forming.

With the addition of available in-feed and out-feed equipment, an orbital forge can be made into a fully automatic system. The hydraulic tool changing clamps enable an operator to change tools from one production run to another in less than 30 minutes.

It appears that orbital cold forging is still in its infancy with regards to industry awareness, but it is destined to gain wide acceptance in the near future.

18

ROLL FORMING

PROCESS DESCRIPTION

Roll forming is a high production process of forming metal parts from coiled, strip, or sheet stock. This is done by feeding the materials longitudinally through successive pairs of rolls, each pair progressively doing its particular forming assignment until the desired cross section is produced. On its way through the rolls, the metal may be hole punched, mitered, notched, embossed, swedged, and/or edge rolled.

Much press or brake work can be done at a substantial savings by roll forming. Roll forming has the additional advantage of being able to feed two different materials, one on top of the other, to form a single product. In this way, desirable properties such as ductility, corrosion protection, tensile strength, and superior finish can be obtained.

Most metals can be roll formed, including anodized, prepainted, and plated materials. Thicknesses ranging from 0.005″ to 0.188″ can be fabricated into desired configurations. The roll forming process produces high quality products; and it produces consistent tolerances on both light and heavy gauge material, also holding shapes and dimensions consistently. Finished parts have an excellent appearance with no die marks, even on precoated or anodized materials.

Applications are increasing steadily because designers are learning about the opportunities roll forming presents. At this time, nearly every industry in the country makes use of the process. The "custom roll former" has had the experience necessary to help you change from other manufacturing processes to roll forming, or to help you design new parts suitable for the process.

CURVING

Roll formed shapes have uniform cross sections, which enables them to be easily bent. Shapes can be curved to uniform radii without wrinkles and without disturbing a prefinished surface. Designers should know that curves can be rolled into a continuous helix and then cut up into sections. They should also know that material elongation is a characteristic which must be considered to prevent wrinkles. A material with a large elongation is going to bend more easily.

PREPIERCING

Prepiercing means punching a pattern of holes while the metal is in the flat strip before forming. It is a continuous part of the same operation of roll forming, but since it is a prelude to that roll forming, and it eliminates an additional and separate handling, it saves money.

POSTPIERCING

Postpiercing is punching a pattern of holes after the forming is completed. But, again, it is part of a sequence of operations, all of which are performed without handling. Sometimes this is preferred over prepiercing because accuracy from the edge of the part would be greater.

FORMING IN LINE

Tabs, stops, raised areas, and welding projections can be formed while the parts are being rolled. This not only improves accuracy but also saves money. Certain operations, which are normally secondary, can be eliminated by incorporation into the roll forming process. The material should be as ductile as possible because that permits crisp design, sharper corners, and easier bending. Bend radii specified by the mill should be followed when high strength alloy steels, heat resistant steels, titanium, etc., are used.

GUIDES FOR ECONOMY

There are a number of design hints for those considering roll forming.

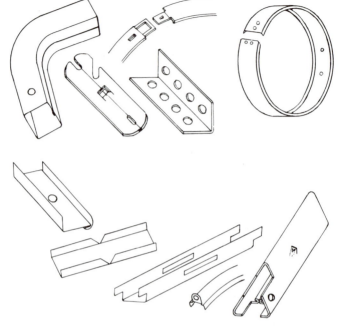

Fig. 18.1. Examples of products produced by roll forming.

1. The depth of bend (channel) should not be too deep for its width.

2. If stiffness has to be increased, it can be done by design. Pressing ribs in flat sections or folding material over to double its thickness will increase stiffness.

3. If flat, wide areas are needed at edges, they can be formed by using stiffening ribs.

4. The leg of a channel or angle should always be longer than three times metal thickness. This also applies when hemming or bending metal back on itself.

5. When using a piercing pattern which must be located a specific distance from the end, try to make this distance between ½" and 4".

Fig. 18.1 shows some applications of these design considerations.

TOLERANCES

The tolerances given here are only guidelines; if tighter tolerances are required, discuss them with your roll former. Dimensions in certain areas can be held more closely than in others.

1. Cross sections can be held to ±0.010″ and angles to ±1°.
2. Straightness tolerances (bow or camber): 0.015″ maximum deviation per foot of length.
3. Twist tolerance is ½° maximum deviation per foot of length.
4. Length tolerance for parts 0.015″–0.025″ thick:
   ±0.020″ on parts up to 36″ long
   ±0.047″ on parts 36–96″
   ±0.093″ on parts 96–144″.
5. Length tolerance for parts 0.026″ thick and greater:
   ±0.015″ on parts up to 36″ long
   ±0.030″ on parts 36–96″
   ±0.060″ on parts 96–144″.

More Design Hints — These suggestions may be helpful.

1. Use the largest bend radius permissible whenever possible. An inside bend radius of less than metal thickness will use more power and decrease roll life.
2. Design the part to be as symmetrical as possible to preclude twist in the completed part.
3. Do not position holes, slots, and notches too close to a bendline to avoid distortion.
4. To maximize reduction of tool and part cost, do not request tolerances which are tighter than required.
5. The outside leg on panel stock will be wavy when rolled straight. Forming a slight bend in the leg will tend to keep it flat.
6. The flat 180° hem will be wavy. By forming a teardrop hem, the edge will stay flat.

7. A 90° leg on panels adds another step and might mark the part. Forming a 75° leg reduces the manufacturing steps and will not mark the part.

8. Avoid sharp inside radii since they are difficult to form without marking the outside radius or cracking paint on prepainted metal. A larger radius reduces the problem.

9. Roll breakage might occur when the design calls for a narrow slot. Breakage can be reduced by widening the slot.

10. When there is a roll form at one end and a wide flat at the other, put a groove near the end or put a leg on the end to maintain flatness and straightness.

11. If you cannot control a wide sweeping radius, put a bend on each end to keep the metal straight.

EXPENSE

Normally, the roll former wants a minimum order of 10,000 feet; however, he would probably run a 5000 foot prototype order. Rolls are inexpensive tools to make. Very often, the roll former can find the rolls you need in his tool room of stock dies. Dies are often available for all sizes. They can fabricate round or rectangular tubing, angles, or channels. The tubing wall thickness runs between 0.009" and 0.090". Angles run between 0.009" and 0.188" in wall thickness. The wall thickness of channels runs between 0.009" and 0.125".

19

THERMAL ENERGY DEBURRING

INTRODUCTION

Over the years, industry has made great strides using modern machining methods which have increased productivity and improved the quality of manufactured parts. However, most of the attention has focused on primary machining methods, while finishing and deburring parts has been largely ignored. Today, 10% (and sometimes much more) of total manufacturing costs are spent on manual deburring.

All designers should consider deburring because it is so often a problem. Some machine shops insist that their parts are burr free. Burrs may be tolerable, but they are generally not desirable. Very often, groups of operators are put to work hand deburring parts for aesthetic reasons or because the parts will not function properly with burrs present.

When the engineer determines that deburring must be done, there really are alternatives. First of all, he can consider various machining methods which might preclude the necessity of deburring. Then he can choose between four machine methods of deburring, which shall all be reviewed in this book.

Around 1975, a commercial process was developed to offer industry an alternative to costly hand deburring—the process became known as the Thermal Energy Method of deburring (TEM). It is the fastest method in existence. Not counting loading and unloading time, the actual deburring time is less than 30 milliseconds.

THE PROCESS

The manufactured parts, with burrs, are placed in a thick-walled chamber which is sealed and pressurized with a mixture of oxygen and natural gas. (The ratio of oxygen to gas is $2\frac{1}{2} : 1$.)

The chamber is closed and sealed with a toggle mechanism exerting a force of 250 tons. The gas mixture fills each nook and cranny of all parts in the chamber (even blind and intersecting holes); and the combustible mixture is ignited by a 30,000 volt spark which creates a 6000° F heat wave. (See Fig. 19.1.) In a few milliseconds, the fuel is burned out. But since most burrs exhibit a high surface area-to-mass relationship, the burr cannot transfer heat to the main part fast enough to prevent its bursting into flames. Thus, the burr becomes a source of fuel and will continue to vaporize until the heat is transferred to the part itself. As the heat moves into the part, the flame temperature drops until it extinguishes itself. By this time, all burrs, chips, and contaminants have been vaporized.

In the process of vaporizing, the burrs become oxides of the metal being processed: aluminum oxide from aluminum parts and iron oxide from steel parts. The oxide settles on the parts

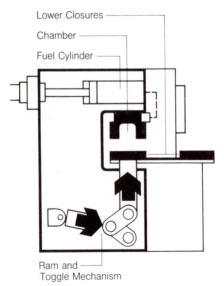

Fig. 19.1. The thermal energy method (TEM) unit. (Courtesy of Surftran Corporation.)

as a loose powdery residue. Although the surface will appear discolored, it has not been oxidized and the discoloration will wash off with a suitable cleaner. A cleaning step can be avoided if the deburred parts are to be heat treated, bright dipped, black oxided, anodized, or otherwise treated or plated.

If postcleaning should be necessary to make the parts more presentable, cleaning equipment is available from the same company which manufactures the TEM machine. As a matter of fact, whenever you discuss deburring with a machine builder, you should also discuss cleaning if it is going to be required. The complete finished part is your goal, and if cleaning is significant, it should be part of the manufacturing strategy. This is especially true when the cost of cleaning equipment equals the cost of the deburring machine.

The list of industries which are taking advantage of TEM is growing annually. The most recent technical advance was the ability to process plastics as well as metals. TEM is really another manufacturing tool to be used when the situation warrants. The design engineer should consider the deburring problem before completing his design. Recently, the author watched a crew of six operators hand deburring rocket tails. The job took an average of thirty minutes per part, and it left unwanted scratches on the surface. Fifty percent of the cost of that part was deburring.

A most unique aspect of TEM is that the deburring media is gas. Normally, media refers to an abrasive used in the deburring process. Some parts require fixtures to support them against the shock wave; fixturing may also be necessary to heat sink very thin parts. Small parts can occasionally be batch processed, especially if they only have internal burrs. In actual production, most parts are deburred without fixtures. When parts must remain scratch free, or if diecast parts which are not heavily ribbed are being processed, those parts should be held or supported. There are no rules about what should be fixtured. Experience is the best teacher.

TEM will not change any dimensions, surface finish, or physical properties of the part, providing fixtures are used where needed. That is because the parts are exposed to great heat only for milliseconds and the exposed parts seldom see a temperature higher then a few hundred degrees Fahrenheit. Threads are not affected by the heat because their wide roots transfer heat quickly.

LIMITATIONS OF TEM

Despite all of the advantages of TEM, there are some drawbacks. If the parts go into the deburring chamber with oil on them, the oil will burn into the workpieces and leave a carbon smut on them which will be difficult to remove. In addition, the oil might prevent gas from washing the surface properly. Before entering the chamber, all parts must be free of oil. Also, blind and tapped holes must be free of compacted chips. Unless the gas can surround all chips, it is unlikely that they will vaporize completely. The final problem has to do with consistent radii. Getting consistent radii on aluminum and stainless is difficult to achieve. There is another deburring method called "Abrasive Flow Deburring" (described in the next chapter) which provides uniform radii when that is vital.

TEM is effective on most engineering materials, although it is more suitable on some than others. As previously mentioned, the burr must absorb heat, reaching a high enough temperature to be oxidized. These two critical steps are difficult to achieve if the material has a high heat transfer coefficient. Nevertheless, there are companies with sufficient knowledge and experience to process many metals, even those difficult to deburr like copper and aluminum. See Fig. 19.2 for examples.

Another significant limitation is size. The newest machines cannot handle anything larger than 10" diameter by 30" long. However, if a special size chamber were required, it could be built — for a price.

Postdeburring cleaning is another consideration. Not all burrs are the same size; but most are close enough in size not to present a problem. Occasionally, a very large burr is not completely oxidized. That burr will take the form of a small molten particle, like weld spatter. That partially oxidized burr must be removed mechanically.

APPLICATIONS FOR THERMAL DEBURRING

The following are some benefits of using TEM:

Fixed manufacturing costs with no variances;
Elimination of costly and time consuming manual deburring;
Assurance that all burrs are consistently removed; and
Increased quality levels and reliability.

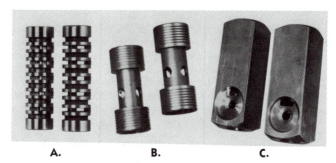

A.   B.   C.

Fig. 19.2. Parts deburred with TEM. **A.** I.D. windows on these aluminum automatic transmission components (3.5″ × 1″) were deburred at 720 parts per hour. **B.** All loose burrs on these valve spools (2″ × 0.75″) were removed at 1440 parts per hour. **C.** Intersecting holes on these steel hydraulic valve bodies (4″ × 1.5″) were deburred at 360 parts per hour. (Courtesy of Surftran Corporation.)

Specific industries have benefited from TEM; e.g., screw machine products are very often small, with many holes and sometimes cross holes. Whatever the material—brass, aluminum, steel, stainless steel, or plastic—screw machine parts can readily be deburred by TEM.

The die casting industry produces many small parts, in both zinc and aluminum, which require deburring. Just consider the cross holes in a carburetor; TEM was created for parts like carburetors.

In the fluid power industry—both hydraulics and pneumatics—it is crucial to remove all burrs. A loose burr can wreak havoc in fluid power. In these "swiss cheese" manifolds, complete burr removal is guaranteed by TEM. TEM also deburrs spools, cartridges, valve bodies, and other components. Cast iron parts are particularly successful with TEM, not only for deburring, but because core sand is blasted out leaving clean, surfaces.

Consider car door lock plugs; before TEM, these plugs were manually deburred by the thousand every week at high cost. They can now be batch deburred by TEM at a machining rate of about 400 per minute.

There are many places in the aircraft industry where TEM comes in very handy. Sometimes edges are manually broken and then followed by a TEM treatment. In fuel pump housings, hidden trash can be eliminated by TEM. TEM also does an excellent job on jet engine vanes.

Recently, TEM has been used to decore sand castings. Apparently, TEM is able to melt binders out of foundry sand which allows sand to be easily poured out of intricate castings.

EXPENSE

The cost of TEM equipment is high; small machines are priced at $100,000 and up, and one can easily spend $250,000 for TEM equipment and surface cleaning machines. Maintenance and cycle cost is in the neighborhood of 10¢ per cycle, so the actual operation of TEM equipment is very inexpensive. Therefore, any time that the machine can be kept in meaningful operation, it will earn you money. There used to be a company in Southern California that employed dozens of people to deburr, file, and clean components for the aeronautics industry. Three years ago, when they were visited, they had automated the operation with specialized deburring machines, and few personnel were visible.

There are over 200 TEM machines in current operation. No matter where you reside, you will not be too far from some company anxious to do job shopping with TEM machines. Job shopping is also done by Surftran, the company which introduced TEM. The company address is: Surftran, Robert Bosch Corp., 30250 Stephenson Highway, Madison Heights, MI 48071.

20

ABRASIVE FLOW DEBURRING

PROCESS DESCRIPTION

Abrasive flow machining (AFM) finishes surfaces and edges by forcing a flowable abrasive media through or across the workpiece. Abrasion occurs only where the flow of media is restricted, otherwise the abrasive has no effect. The process works on many surfaces or selected passages simultaneously, reaching even seemingly inaccessible cross holes and interior areas.

Several parts can be abraded at one time, thus yielding rates of production of hundreds per hour. A variety of finishes can be produced at the same time by altering process parameters. In production applications, tooling is designed to be loaded and changed quickly. In current industrial practice, the last remaining high cost area is part finishing or deburring. Unless one of the new machining methods for deburring is utilized, finishing remains a labor intensive, uncontrollable area.

Proper finishing of surfaces and edges affects more than simply the feel or appearance of a part. Performance can be dramatically improved by the correct finish. AFM is another competitor to Thermal Energy Deburring (TEM) and Electrochemical Deburring (ECD) in the area of product finishing by machine. With today's focus on total automation with machine tools, AFM, TEM, and ECD all offer flexibility as an integral part of the complete manufacturing cycle.

Abrasive flow machining is used in many applications involving deburring, polishing, and edge radiusing. Advances in both tool design and media formulation have established AFM as a means of satisfying difficult manufacturing requirements.

PROCESS FUNDAMENTALS

In AFM, two opposed cylinders extrude an abrasive media back and forth through passages formed by the tooling and the workpiece. Since the abrasive is active only where passage is restricted, the tooling is very important. One throughput can deburr one section of the part, polish another section, and establish a definite, repeatable edge radius. Lapping can be done if one allows the media to gently hone a surface. Fig. 20.1 shows the machine which does this.

The process handles soft aluminum, stainless steel, tough nickel alloys, or even hard materials like carbides and ceramics, and AFM achieves a wide range of predictable results. If the process objective is uniform polishing, the media should maintain a uniform flow rate. If the objective is deburring or edge radiusing, the flow should be increased where you want the ac-

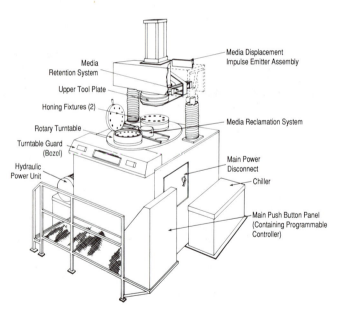

Fig. 20.1. An Extrude Hone orbital abrading machine. (Courtesy of Extrude Hone Corporation.)

Fig. 20.2. Extruding media passing through a workpiece. (Courtesy of Extrude Hone Corporation.)

tivity to occur. This change in flow is a direct result of tool configuration, the media, and the machine setting. See Fig. 20.2.

The machine controls extrusion pressure, which is adjustable from 100 to 3200 psi, with flow rates exceeding 100 gallons per minute. The volume of flow depends on the displacement of each cylinder stroke and the total number of strokes used to complete the job.

Control systems can be added to monitor and control process parameters such as media temperature, viscosity, flow speed, media feed, media cooling, and load–unload stations. Such automated systems can process thousands of parts per shift.

TOOLING

The tooling holds the workpiece in position and directs the flow of media to the appropriate areas. In this latter function, it may also restrict flow where abrasion is desired, or it may completely block the flow through areas where no change is desired. Many AFM applications require only simple tooling.

Dies generally require no tooling; the die passages themselves provide the necessary restrictions. To process external surfaces, the space between the inside of the fixture and the outside of the part will be restricted as needed by tooling.

MEDIA

The AFM medium is composed of a semisolid carrier and a concentration of abrasive grains. Specific results can be achieved by varying the abrasive grain size, type, and concentration as well as the viscosity of the carrier. Low viscosity media are used for radiusing edges and smoothing small passages, while higher viscosity media are appropriate for smoothing the walls of large passages. When abrasives enter a restrictive passage, the viscosity of the media temporarily rises, thus holding the abrasive grains rigidly in place. When in this rigid condition, the media abrades the passage and then softens to its original state, producing little or no abrasion.

Boron carbide, aluminum oxide, and diamond may be used as abrasive media, but most jobs use silicon carbide. Particle size ranges from 0.0002″ to 0.060″. As expected, the larger grains remove stock at a faster rate, while the smaller size grains provide finer finishes and access to small holes.

The effective life of the media depends on several factors starting with the initial batch of media: the abrasive size and type, the flow speed, and the part configuration. During the AFD process, the abrasive grains break and become less effective, and the abraded material mixes with and dilutes the media; however, this does not happen quickly. A typical machine load of media can be used for weeks to process thousands of parts before replacement. Air nozzles or vacuum can remove media from internal passages, and removal can be completed by means of a solvent wash.

Fig. 20.3. Investment cast compressor wheels can be polished to maximum efficiency. (Courtesy of Extrude Hone Corporation.)

PROCESS APPLICATIONS

Precision, flexibility, and consistency are available through AFM. There is a large number of applications for AFM in the aerospace and automotive industries, production, and all kinds of die finishing. Diverse applications, including surgical implants and centrifugal pumps, have materialized. Fig. 20.3 shows another application.

The process was originally devised for critical deburring of valve spools and bodies. It performed so well deburring edges that additional tasks were found for the process; e.g., any parts with internal crossed holes can be quickly and consistently handled. The process is equally applicable to small (0.060″ diameter) gears and passages several inches wide. Abrasive flow deburring/machining is an established process. The Extrude Hone Corp. of Irwin, PA, has had a great deal of experience with it, and could be of tremendous assistance to any company interested in either purchasing AFM equipment or using the jobbing services of Extrude Hone.

21

ELECTROCHEMICAL DEBURRING

PROCESS DESCRIPTION

Electrochemical deburring (ECD) is a deburring process which uses electrical energy to remove burrs in a very localized area, as opposed to thermal energy machining which provides general deburring. The part to be deburred is placed in a non-metallic fixture which positions an electrode in close proximity to the burrs. The workpiece is charged positively (anode), the electrode is charged negatively (cathode), and an electrolytic solution is directed under pressure to the gap between the electrode and the burr. This flow of electrolyte precedes the application of the current in order to flush out any loose chips which probably would cause a short in the system that could damage the part, the tooling, or the equipment. As the burr dissolves, a very controlled radius is formed. The process is consistent from part to part.

The process always requires fixturing to establish the anode–cathode relationship. A typical fixture consists of a plastic locator which holds the part and insulates (masks) areas of the part which do not require ECD. The fixture also positions a highly conductive electrode, designed with a contour that conforms to the desired dimensions of the area to be deburred. The locator and electrode direct the flow of electrolyte. The variables of voltage, current, electrolyte flow, and cycle time provide precise control of the ECD process. The process is depicted in Fig. 21.1.

APPLICATIONS OF ECD

ECD is effective on all electrically conductive materials (copper alloys and stainless steel are both good materials for the process). The benefits of using ECD include:

Elimination of costly hand deburring,

Fixed manufacturing costs,

Assurance that all burrs are consistently removed,

Increased quality and reliability, and

Radius generated during ECD is controllable.

This last advantage solves functional problems such as removing sharp edges from the ID of valve bodies where cross holes intersect; this is also applicable to ports in hydraulic components. In hydraulics, there are many occasions when seals must slide over sharp edges both at assembly and during operation. The slight radius resulting from the ECD operation precludes the possibility of harming elastomer seals during this assembly.

In the instrument industry, burrs created on delicate gears during the hobbing operation must be removed without damage; ECD does an excellent job here. In fact, ECD will actually improve the surface of these gears. In both the ordnance and aeronautic industries, ECD and thermal energy machining are sometimes combined when specific finishes are required, and each process is used where it excels.

The ordnance industry has an exacting specification. The ordnance component must work once, upon demand; there are no second chances. Many of the components function only because of the tremendous forces, accelerations, and RPM's created by the power of an explosion. This same power can cause malfunctions if the sliding, indexing, or rotating surfaces and edges are not properly deburred. These same forces can dislodge a single, tiny burr and jam a critical movement. The manufacturing engineer enjoys maximum assurance that all burrs are removed when he knows that ECD and TEM have been part of the manufacturing cycle. See Fig. 21.2.

ELECTROLYTIC ACTION

During the ECD process, there is no contact between the workpiece and the "tool." Thus, there is no metallurgical change as a result of the electrochemical process. In grinding or polishing to remove burrs, the workpiece is exposed to mechanical and thermal stresses.

Faraday's Law of electrolysis dictates how metal is removed by ECD. The amount removed is proportional to the product of

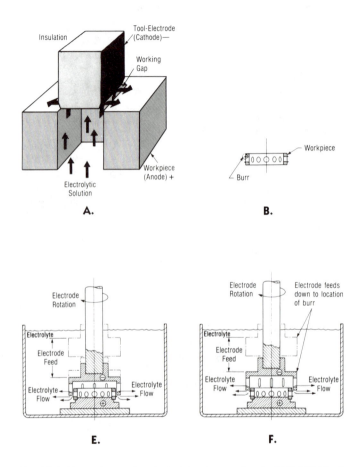

Fig. 21.1. How electrochemical deburring works. **A.** DEBURRING of a workpiece by electrolytic means relies upon "deplating" the anodically connected workpiece, using a cathodically connected tool, both immersed in electrolyte such as salt water. **B.** WORKPIECE with a burr. **C.** WORKPIECE mounted on anode connection in a tank of electrolyte. **D.** CYLINDRICAL brass tool has slots to cause turbulence, and is connected to negative lead from

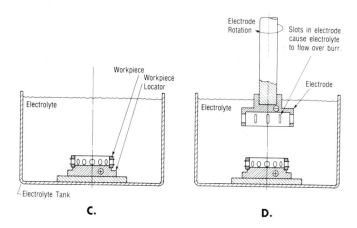

C.   **D.**

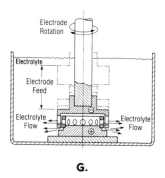

G.

a direct-current source. **E.** CATHODE-TOOL is lowered while rotating, advancing toward workpiece. **F.** TURBULENCE created by rotating tool flushes the area being deburred with fresh electrolyte as tool advances on part. **G.** TOOL CONTINUES DOWNWARD as burr is removed. Electrode rotation is reversed frequently to create more turbulence. (Courtesy of Surftran Corporation)

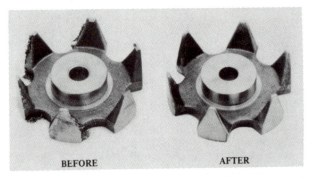

BEFORE   **AFTER**

Fig. 21.2. An alternator pole piece half deburred by the electrochemical process. (Courtesy of Electrogenics Company.)

time and current. Recently, the process has been simplified by the use of rotating electrodes that create turbulent flow of the electrolyte, which accelerates the deburring process. The second advantage that this creates is that it allows the use of standard tools more often rather than requiring custom tools for each job.

When using a rotary electrode, the workpiece is positioned on a holding device which is connected to the positive lead from a DC source. The cathodically connected tool can be a cylindrically shaped, brass, rotating tool slightly larger than the part to be finished. There should be some vertically machined holes in the tool which help improve the flow of electrolyte. The tool is often made from brass pipe.

In operation, the tool is fed downward toward the workpiece which is immersed in the electrolyte. The rotating electrode creates turbulence of electrolyte around areas to be deburred (the spindle is reversed frequently to increase the turbulence). After a cycle of about 30–45 seconds, the spindle retracts and the part is removed.

ECD and TEM are used in nearly all of the same industries. The decision of which process to use depends on the requirements of the part and the capabilities of the processes. TEM is mainly used for deburring. ECD is used for deburring also, but it will round off edges as much as necessary and can be used for polishing to some degree.

Now where does this leave Abrasive Flow Machining? AFM deburrs, it smooths edges, and it certainly is used to polish dies. Apparently there are areas where each process excels, and yet it is possible to achieve the goal by using more than one of these processes. Therefore, you should evaluate each process just as you would any machine tool before purchasing because that is what these "deburring machines" are. You have got to determine how many hours of work you could obtain from each.

There are two good sources of information on these processes, and you should take advantage of both: the machine tool manufacturer; and those who have purchased these machines.

22

ELECTROCHEMICAL MACHINING

PROCESS DESCRIPTION

Electrochemical machining (ECM) is one of those nontraditional machining processes which is finding more applications as equipment improves and manufacturing engineers learn more about it. First, the aerospace industry found it very useful; then applications were found in the automotive, appliance, machinery, military hardware, and even medical implants industries.

Electrochemical machining is a controlled, rapid metal removal process with virtually no tool wear. Unlike conventional processes, ECM removes metal atom by atom. It can be regarded as a solution to a variety of metal removal problems such as cavity sinking, radiusing, contour machining, and machining helices (rifle barrels).

ECM equipment for production is usually built to suit a specific machining task, although sometimes a standard machine tool will be modified. A complete installation consists of the machine, the electrical system, the electrolyte system, and the electrodes. The type of machine depends on the workpieces. Should the machine be single-station or twin-station? Should it be vertical or horizontal? Should the machine be automated or have an indexing table or some other special equipment?

Electrochemical machining uses electricity, chemistry, and some mechanical components usually arranged as depicted schematically in Fig. 22.1. The cathode tool (A) is shaped to provide the form desired in the workpiece (B) with appropriate compensation for overcut. The Anocut power unit supplies 5–20 volts DC with negative charge through cables to the tool, and positive charge to the workpiece. An electrolyte (electrically conductive solution) is pumped under high pressure between

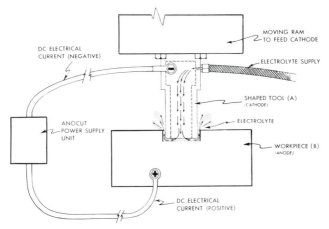

Fig. 22.1. The ECM process. Straight flow tooling shown. (Courtesy of Anocut Inc.)

the tool and the work. At the same time, a mechanically driven ram feeds the tool at a constant rate, preset by the operator, into the workpiece to machine the desired shape.

This is a good spot to explain some differences and similarities between the ECM process and electroplating, electric discharge machining, and chemical milling.

Electroplating is, of course, the commonly used method of finishing metal (sometimes nonmetal) parts; it generally adds a thin coating of a stainless or hard material to a part to protect it against corrosion. EDM is mainly used for tooling purposes such as sinking cavities or machining hardened steel. Electrochemical milling is somewhat similar to ECD in that both processes are the reverse of electroplating. Instead of adding metal to a part, these two processes deplate. The milling is done to very thin conductive metals, and is generally used to remove just a few thousandths of an inch of metal. A photo-etching step prepares the surface very much like a printed circuit board, and then the unprotected (unmasked) areas are deplated.

Tooling for ECM must be designed so that all areas of the tool are properly irrigated and the flow across these areas is smooth and even. Improper flow will cause poor surface finish,

striations, and unreliable operation. Tooling which is not designed correctly to supply sufficient flow can damage the electrode or workpiece due to sparking.

The most familiar use of Faraday's Law is electroplating.
ECM is an application of Faraday's Law in reverse of electroplating — controlled metal removal. Faraday's Law governs the
rate of metal removal, assisted by control of the space between
anode and cathode and the rate at which the ram is fed. All
common metals are removed at about the same rate, although
the exact rate is determined by the electrochemical equivalents.
For estimating purposes, it is convenient to assume that 10,000
amperes will remove 1 cubic inch of metal per minute. The tool
never touches the workpiece, so there is never any tool wear or
damage from heat or sparking.

STRAIGHT FLOW TOOLING

Previously, the common method of providing for electrolyte
flow was to design electrodes with holes or slots in them
through which electrolyte was pumped. Electrolyte normally
passed across the face of the tool and exited along the outside
of the tool between it and the wall of the cavity being machined.
For example, to machine a round hole, a hollow tube becomes
the electrode. Electrolyte is pumped down the inside of the tube
and up the walls of the machined hole as illustrated in Fig. 22.1.

There are some disadvantages to this general method.

1. The freely exiting electrolyte creates striations and a
   poor finish.
2. The flow from inside out thins and sometimes is insufficient to provide a smooth finish.
3. Complex electrodes can end up with areas starved for
   electrolyte.
4. The electrolyte tends to spatter in this type of flow.

REVERSE FLOW TOOLING

Some of the problems with straight flow tooling can be overcome by reversing the flow of electrolyte so that the electrolyte
is introduced between the electrode and the work by means of
a dam or supply chamber. The electrolyte enters the work zone

around the periphery of the tool. The exit passage could be holes or slots in the electrode face; and each exit passage is connected by a hose to the electrolyte tank.

This procedure leads to:

1. A better finish,
2. A uniform and predictable overcut,
3. Freedom from sparking,
4. Cleaner operation, no spatter, and
5. A direction of flow which prevents undesirable erosion.

DESIGN CHARACTERISTICS OF REVERSE FLOW TOOLING

Fig. 22.2 shows a simple form of reverse flow tooling. Clamps are used to hold the dam down against the workpiece. (It is customary to provide more than one inlet connection in order to ensure equal distribution of electrolyte.) The dam can be made of brass, stainless steel, or green glass. If the dam is made of metal, a plate of green glass must be secured to the side of the work to insulate it.

Fig. 22.3 shows an improved reverse flow tool. Clamps are eliminated, and the dam is constructed so that the pressurized electrolyte seats it against the workpiece.

If an electrode has an irregular outline, its shank can still be made straight or round. This is easier to build. To prevent stray etching of the workpiece, a plate, having a cutout in the shape of the electrode, can be mounted on the end of the dam. The cutout should be about 0.030" larger than the electrode that fits through it. This is to ensure a uniform electrolyte flow around the electrode.

The green glass plate, which has been cut to conform to the electrode's shape, does not have to be very thick because it transmits the hydrostatic force directly to the workpiece. However, the sidewalls of the dam must be strong enough to withstand substantial hydrostatic forces.

ECM ELECTROLYTE SYSTEM

Electrolyte is pumped through the working gap, and it removes metal deplated from the workpiece and heat generated

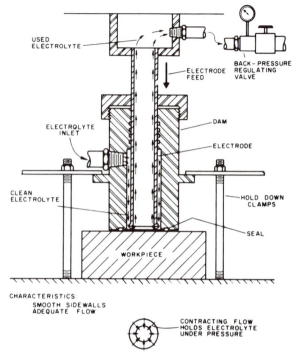

Fig. 22.2. Reverse flow tooling. (Courtesy of Anocut Inc.)

during the electrolytic action. The electrolyte is stored in a tank and filtered regularly; the pressure is regulated and controlled. In order to maintain the electrolyte machining characteristics constant, the pH value, the concentration, and temperature are regulated to preset values by means of three individual regulating units.

The dissolved metal precipitates in the electrolyte as metal hydroxide and can be removed by gravitational forces, by means of either a centrifuge or a settling tank. Normally, the wet sludge is pressed into a 50% solid block. Before entering the filter press, the wet sludge can be treated to a neutral pH to avoid environmental pollution.

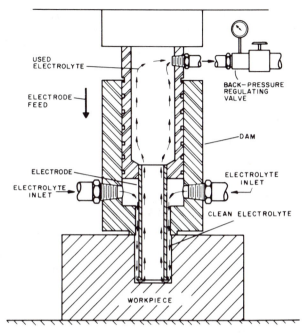

USED ELECTROLYTE

BACK-PRESSURE REGULATING VALVE

ELECTRODE FEED

DAM

ELECTRODE

ELECTROLYTE INLET

ELECTROLYTE INLET

CLEAN ELECTROLYTE

WORKPIECE

Fig. 22.3. Improved reverse-flow tool. (Courtesy of Anocut Inc.)

ECM PRODUCTION APPLICATIONS

Turbine wheels are electrochemically machined blade by blade. For this job, the electrode is a metal plate with a hole of similar cross section to that of the blade to be machined. ECMing speed for this assignment is between 0.2 and 0.3 inch per minute.

Constant twist or gain twist rifling can be ECM broached into weapons like pistols, rifles, or even cannons at rates up to 20 inches per minute.

Profiles of irregular shapes in high volume can be done economically. Six or more parts can be processed simultaneously, using a cassette to hold them in position. Irregular cavities and shapes which are difficult for conventional machining can be easily processed by ECM. Machines equipped with integral rotary tables are used to make a variety of cavity configurations in engine casings. As many as six different cathodes can be used for multiple machining around the casing OD. Relatively large cavities can be economically machined by ECM. Some shapes can be processed in $\frac{1}{10}$th the time required by EDM.

Gap protection devices for ECM equipment consist of five different detectors, which respond to short circuit, turbulence, passivation, contact, and overcurrent.

A machine tool capable of cutting deep splines into long tubes with an accuracy and speed not previously possible is now available. The technique uses the principle of reverse flow ECM, employing a unique cutting element design.

Specimens for metal fatigue testing and tensile testing can now be machined quickly and expertly on a special specimen grinder. Special electrochemical grinders are also available. Although called a grinder, 90% of the metal stock is electrochemically deplated; only 10% of the material is ground off. This process provides burr-free smooth finishes. It is good for thin, fragile parts and wherever cool, stress-free grinding is useful. The workpiece is positively charged with a low voltage DC current. The conductive grinding wheel forms the negative leg of the process. The high rate of stock removal is due to formation of metal oxides by high DC current flow. These oxides are then removed by the abrasive grinding part of the process.

In conclusion, ECM is here to stay. It behooves manufacturing engineers to consider its cost saving possibilities when preparing for new production or reviewing old.

23

COMPUTER INTEGRATED MANUFACTURING

HISTORY

Computer integrated manufacturing (CIM) really started with Eli Whitney's revolutionary approach to manufacturing back in 1798. He introduced the idea of making interchangeable parts for ease of assembly. Then, around 1970, we witnessed the introduction of computer aided drafting (CAD). This was a giant step forward. CAD databases contain complete part definition including design, dimensions, and tolerancing data. If we could drive manufacturing and inspection data directly from this CAD database, the complete part manufacturing process could be more successfully automated.

As assemblies became more complex, industry needed a better method to describe the shape and dimensions of parts and how they fit together. This is not to belittle the very significant contribution of computer aided manufacturing (CAM). That system accelerates drafting and research and development and is absolutely a modern miracle. But, still, success toward acquiring a complete CIM system has been limited.

Without a doubt, automated tools have been a great advance for manufacturing. However, getting various types of equipment to work together has been a different story. We have needed some way to use CNC machine tools, coordinate measuring machines (CMM), and other automated equipment in a related manner. Modern CIM software provides a direct and simple linking system for all this equipment.

PROCESS DESCRIPTION

Complex fixturing is no longer required since machining and inspection paths can now be oriented by software as the parts are positioned. This improves productivity considerably. The bottom line is increased value out of tremendously expensive investments.

When a part is placed in a machine tool, the traditional manufacturing technique is to precisely fixture the part within the X, Y, and Z planes; any inaccuracy in positioning will adversely affect subsequent production. Computer software simplifies these part setups by providing the intelligence for a machine tool probe to automatically determine the part's exact orientation to its own X, Y, and Z planes via data referenced to part features rather than the fixture itself.

If you want to know whether the parts just made will fit together in an assembly, they must be inspected. There are only three methods of verifying the fit: hard gauges or gauging fixtures, coordinate measuring machines, and special inspection machines.

1. Hard gauges are normally used in high volume production. Sometimes two gauges are needed to check the maximum and minimum tolerances. This type of gauging is easy to use. For instance, the worker simply places the part in the gauge to see if it is within tolerance. As long as the part dimensions do not change, the gauge cost is not high. The disadvantage of these gauges is their inflexibility. Scrapping or reworking gauges can be expensive. Then, of course, the time to build can be a month or more.

2. Coordinate measuring machines are precise measuring instruments which can provide an electronic, digital readout that can be understood by a computer. They work slowly compared to hard gauges. They are also very expensive — starting at $25,000 for a small manual machine, they can cost as much as $1,000,000 for large, fully automatic machines. Despite these disadvantages, CMM's are growing in popularity. In fact, since some models started to be produced with ceramic material used for the structural parts, these CMM's are being placed in open shops so that the machine operators can inspect alongside the production machines. CMM's are flexible and can be reprogrammed easily.

3. Special inspection equipment has an important role in several large industries, especially the automotive. Pistons, camshafts, crankshafts, and gears are inspected automatically. Lasers and other vision systems are often used with the special equipment.

COORDINATE MEASURING MACHINES

CMM's are now designed for use in the factory environment. Shortly, manufacturers are going to want the advantages of superior, real-time communication between shop and manufacturing engineering, and between manufacturing engineers and design personnel.

It has been difficult to link CAD to CMM's. Although machine tools and inspection programs appear to be similar, there are significant differences in the information required to generate both types of programs. NC operations are motion oriented: moving, turning, cutting, starting, stopping, and turning on coolant. The action of a CMM automated program requires other types of steps: select probes, collect and analyze data, and report data.

To understand why the CMM–CAD link is so important, it is helpful to review how the CMM works. In all CMM's, a switch, called the probe, is mounted on a mechanism which allows the probe to move along three perpendicular axes. Sensors continually monitor the position of the probe. When the probe touches a surface of the part being measured, the location of the probe is recorded.

Part of the problem of linking CAD and CMM's is that CMM's from various vendors differ in the way they analyze inspection data, and thus they produce different analyses of the same part. Although each machine will inspect the same points, the internal processing algorithms which process and report the data differ from each other.

CAD users are very interested in CMM's which move their probes by means of servo motors. The CMM's generally are controlled by small computers similar to those which operate numerically controlled machine tools. These CMM's can be programmed to follow a predetermined path and stop at selected points to measure those positions on the part being inspected.

The data gathered this way can be stored and analyzed by

other computer programs. Programs can be prepared which will automatically accept or reject parts by comparing dimensions taken from CAD data with those acquired by the CMM. Data gathered from a quantity of CMM inspections could be used for statistical process control.

Since computers are now less expensive, CMM manufacturers are producing more sophisticated systems. CMM's are available which can store complex programs. Unfortunately, these programs are usually in proprietary language and cannot be transferred from one machine to another. The programs of one company—Valisys—are unusual in that they will run on many different machines.

For computer integrated manufacturing to be fully effective, there has to be a way for manufacturing engineers to communicate with design engineers as well as the shop floor. Some software suppliers have designed software which fills this gap. This software automates the comparison between the product design database and the actual machined part—it permits the part to be inspected before its removal from the machine.

The general software strategy is to complement a CAD/CAM system by providing tools to verify the design process and to integrate functions important to quality control and inspection. In fact, the only step the software does not take is the actual cutting of material on the NC machine. A separate tape is still required for that task, although software will align the part in its X, Y, and Z axes in the same way a part is positioned in a coordinate measuring machine (CMM). And if the part is removed for some reason, it permits the inspection of the part on a CMM automatically, by comparing the part to the product design database. The software actually improves quality control and production management.

THE VALISYS PROGRAM

One of the dynamic software suppliers is the Valisys Corporation which was formed in November 1987 after a few years of development work. Since then, the company has been developing and supporting manufacturing automation software products that validate part designs, use design data to automate inspection processes, and support quality control analysis.

The first decision of the Valisys development team was that a universal design language had to be used to allow design information from various geographical sites to be exchanged. The task was simplified because of the universal acceptance of the ANSI standard Y14.5, which describes geometric dimensioning and tolerancing (GD&T). The symbols used in GD&T provide labels for engineering drawings that not only provide dimensions, but also indicate functional relationships and other characteristics. This standard code prevents misinterpretations.

At this time, it is useful to restate the main problem confronting computer integrated manufacturing: it is the difficulty communicating engineering data from product designers to manufacturing engineering to the shop floor. Now, the Valisys Corporation's software, which is one of the modern manufacturing software systems that improves the flow and accuracy of engineering data in a manufacturing environment, will be described.

The following is a description of nine Valisys software packages: V:Check, V:Gauge, V:Tolerance, V:Path, V:Interface, V:Inspect, V:Qualify, V:Track, and V:Control.

V:CHECK

This package facilitates and validates dimensioning and tolerancing in CAD. The results are greater productivity for designers, fewer design errors, and reduced design review periods. V:Check permits the graphic representation of GD&T symbols to be picked, using standard CAD methods, reducing the designer's keyboard input. It also verifies that the design dimensions and tolerancing are valid, consistent, and in conformance with design standards. If an error is detected, V:Check calls attention to it, describes the error, and prompts for a correction.

Both plus/minus and GD&T are supported in conformance with ANSI standard Y14.5. Because of V:Check's error correction support, it helps designers improve their skills in using GD&T. V:Check performs syntax checking; after the designer has finished dimensioning and assigning GD&T symbols to a CAD drawing, V:Check prompts the designer to designate each symbol he wants to be checked for syntax (for example, its proper use, location, and sequence).

GD&T is an exact language and must be used precisely. Reference data are critical, and symbols must be used in the prescribed sequence. Syntax errors make accurate interpretation of GD&T symbols difficult and unlikely when a number of people have to view the drawings.

When GD&T is understood and used correctly, it reduces guesswork, lowers tool cost, reduces scrap, and provides a common engineering language. A facility with GD&T is significant because government contracts require its use. V:Check (or any similar system) assures design integrity and reduces the number of engineering changes.

V:GAUGE

Working from the dimensioning and tolerancing data generated in the design phase, V:Gauge generates a three-dimensional model of the worst case mating part; that means a software package which is the equivalent of a hard tool is available to inspect parts.

This model, known as a "Softgauge," has more than one use. As a visual tool, it enables a designer to examine all features of his design. Then it serves as the CAD design for a hard tool if and when one is desired. And, finally, the Valisys program V:Path uses it to generate a coordinate-based inspection path.

V:TOLERANCE

This software is used to create worst case software models of parts to make certain the separate parts will assemble properly. Without this ability, parts that are within tolerance, as designed, may still fail to assemble properly.

In addition, V:Tolerance generates the optimum sizes for both clearance and threaded holes. Given the fastener specified by the designer, V:Tolerance generates both the design geometry and drafting text describing those holes.

V:PATH

V:Path creates an inspection path that can be run on a CMM. The program ensures that the inspection process incorporates checks of all critical features. It provides the same level of assurance as a hard gauge.

Inspection path definition with V:Path uses the graphics capabilities of the CAD system and is complete and automatic. The operator can preview the path on the screen and make any changes desired before running the inspection on a CMM. Dimensioning and tolerancing information in the part design is transferred automatically into the inspection process, thus saving time and reducing chances of errors.

V:INTERFACE

This program communicates Valisys functionality directly to the shop floor. It connects many devices and services, including terminals on the floor, machining and inspection equipment and robotic part handlers. In other words, V:Interface ties the shop floor directly into the CAD/CAM database. This leads to more accurate data acquisition on the shop floor because of direct monitoring of production processes. V:Interface brings the full sophisticated inspection capabilities to all inspection devices, i.e., not only CMM's, but CNC machine tool probes in particular, allowing for true, in-process inspection.

V:INSPECT

Using inspection parts generated with V:Path and communicating through V:Interface, this program executes inspections by running CMM's, machine tool probes, and other devices like laser systems. V:Inspect accepts process control information (lot number, serial number, operator, date, etc.), drives the probe to a specified point, takes a measurement, and stores all the measurement data until a printout is requested.

V:QUALIFY

This program creates an as-built model from the measured data generated by V:Inspect, and compares this model to the appropriate "Softgauge" model. If the "Softgauge" for each feature fits the as-built features, then V:Qualify signals that the part passes inspection. If the part fails, V:Qualify decides whether or not the part can be readily reworked. If it can be reworked, V:Qualify illustrates the rework graphically on the model of the as-built part and also provides numerical specifications. If the part cannot be reworked, V:Qualify illustrates the error and provides numerical data to support analysis by a material review board.

In addition, each set of inspection results is archived for later access. This is a benefit in government contract work where traceability is required.

V:TRACK

V:Track is used in the inspection phase to perform statistical process control. It accepts real-time data from the Valisys

V:Qualify module to do this, and also produces control charts from the same data source.

V:CONTROL

This is a programming language for automatically executing a series of Valisys functions on machines, accessible through V:Interface. The user defines a series of steps as a job by using V:Control language. These steps can be any Valisys function: a machine operation, an inspection, or any use of data previously acquired. This job is stored as a named file. To run this job, the operator types the file name at the appropriate prompt; without any further operator action, the steps contained in the job file are then performed automatically.

A single job in V:Control can handle up to 32 machines. The only limitation is that only one command can be executed at a time.

The following is an example of one V:Control job.

1. Load castings from a tray onto a machine tool.
2. Machine several features.
3. Inspect the part with a probe.
4. Load good parts onto one tray and bad parts onto another.
5. Print out inspection results.
6. Stop machine if three bad parts in a row are made.

In-process inspection is now a reality—you can put it to work today. Sophisticated inspections can be done while the part is still fastened to the machine tool. Formerly, industry had two alternatives when machining complex parts: either perform all machining, then inspect, which is risky; or do a few machining steps and then take the part out of the machine to a CMM for inspection. Then, back to the machine again and repeat the cycle. Many hours can be wasted this way. Now it is possible to inspect the part as often as the engineer wishes without disturbing the setup at all. It is an automatic process that saves time and achieves quality results.

Valisys is a multifaceted technology with a broad array of applications available for factory cell automation and statistical process control. In contrast to conventional inspection, which measures deviations in dimensions, Valisys measures ease of assembly. If a company wishes to consider "lights out, factory automation," it should possess a Valisys system or something just as good.

It would be advisable for anyone considering the acquisition of CIM-type software to investigate several suppliers. At this time, the author has found these five competitors to be worthy of further investigation:

1. Cadam Inc., 1935 N. Buena Vista St., Burbank, CA 91504.
2. Computervision, 100 Crosby Dr., Bedford, MA 01730.
3. ICAMP, Inc., 186 Bolton Center Road, Bolton, CT 06043.
4. Intercim Corp., 12217 Nicollet Ave., Minneapolis, MN 55337-1650.
5. Valisys, 2050 Martin Ave., Santa Clara, CA 95050.

24

ADVANCED COMPOSITES

DESCRIPTION

Composites are created by combining two or more materials, a reinforcing element, and a compatible resin matrix in order to obtain specific characteristics. The components do not dissolve into each other, as does sugar in water, but they do act synergistically.

A composite may include many common metals combined with resin and fiber. The term "advanced composites" (AC) designates certain composite materials with properties considerably superior to those originally called advanced composites. Currently, industry defines AC as containing fiber-to-resin ratio greater than 50% fiber, with the fibers having a modulus of elasticity greater than 16,000,000 psi.

There are three types of composites. The first is a particulate based material, formed by the addition of small granular fillers into a binder, which increases stiffness but not strength. The second is whisker/flake filler which does increase strength somewhat due to its higher aspect ratio. The third type is a continuous fiber system which, due to fiber continuity, provides the strength of the high performance fiber as well as increased stiffness.

The major difference between the three composite types is the clear distinction between continuous and discontinuous systems. It is only with the continuous types that both stiffness and strength are fully translated into the composite. In practice, it is very difficult to achieve the theoretical potential of the aspect ratio. Hence, when calculating the physicals of a contemplated composite, one should use a reasonable factor of safety. This knowledge comes only with experience, and depends quite a bit on the workmanship of the technician.

AC has come to mean a resin matrix material reinforced with high-strength high-modulus fibers of carbon, aramid, or boron, and generally fabricated in layers. The four basic areas of composite technology are:

Organic resin matrix composites,

Metal matrix composites,

Carbon-carbon composites,

Ceramic matrix composites.

Organic matrix composites are the most common and least expensive.

THE MATRIX

In a composite, the matrix serves two significant functions: it holds the fibers in place, and it deforms under an applied force and distributes the stress to the high modulus fibrous constituent. For maximum efficiency, the matrix material for a structural fiber composite must have a larger elongation at fracture than the fibers. The matrix must transmit the force to the fibers and change shape as required, placing the majority of the load on the fibers. The matrix also influences corrosion, as well as chemical, thermal, and electric resistance. There are two main classes of polymer resin matrices: thermoset and thermoplastic. The principal thermosets are epoxy, phenolic, bismaleimide, and polyimide. Thermoplastic matrices are many. The matrix material must be carefully matched for compatibility with the fiber material and for application requirements.

Most structural composite parts are produced with thermosetting resin matrix materials. In metal matrix composites, the most frequently used matrix is aluminum, although alloys of titanium, magnesium, and copper are being developed.

FIBER SCIENCE

The term "fiber science" applies to a material tailoring discipline that includes type of fiber, percentage of fiber, and oriented placement of the fiber during the production process. Fibers can run longitudinally (warp) or they can run transversely (weft); there is a new weaving technology which allows fibers to run on a bias.

Since the advent of high-strength high-modulus low density boron fiber, the role of fibers produced by chemical vapor deposition (CVD) has become well established. Boron-aluminum was used for tubular truss members which reinforced the space shuttle structure. However, there are drawbacks. There is a rapid reaction of boron fiber with molten aluminum, and a slow degradation of the mechanical properties of diffusion bonded boron-aluminum at temperatures greater than 900°F. These problems led to the development of the silicon carbide fiber.

SILICON CARBIDE (SiC) FIBER PRODUCTION PROCESS

Continuous SiC fibers are produced in a tubular glass reactor by CVD. The process occurs in two steps on a carbon monofilament substrate which is heated resistively. In step one, pyrolytic graphite about 0.00004″ thick is deposited to smooth the substrate and improve electrical conductivity. In step two, the coated substrate is exposed to silane and hydrogen gases. The former decomposes to form beta silicon carbide continuously on the substrate. See Fig. 24.1.

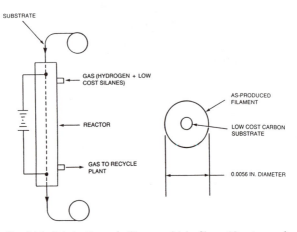

Fig. 24.1. Fabrication of silicon carbide fiber. (Courtesy of Textron Inc.)

The mechanical and physical properties of the SiC filament are:

tensile strength = 500,000 psi
tensile modulus = 60,000,000
density = 0.11 lb/cubic inch
diameter = 0.0056 inch.

HYBRID COMPOSITES

Hybrid refers to the use of various combinations of boron, graphite, aramid, and glass filaments in a thermoset matrix. Hybrids are used to meet diverse design requirements in a cost-effective manner, i.e., better than either advanced or conventional composites.

Some of the advantages of hybrids over conventional composites are balanced strength and stiffness, optimum mechanical properties, thermal distortion stability, reduced weight, improved fatigue resistance, reduced notch sensitivity, improved impact resistance, and optimum cost as related to performance.

The main forms of hybrid composites are as follows.

Interply Hybrids: These consist of plies from two or more different fibers stacked in alternate layers.

Intraply Hybrids: These consist of two or more different fibers mixed in the same ply.

Interply–Intraply Hybrids: These are made up of plies of interply and intraply hybrids stacked in a specific sequence.

Selective Placement: This uses a combination of fibers as needed in any form.

Interply Knitting: This is a form of vertical interply stitching with a polyester or aramid strand. It connects two to five plies and strengthens the composite against interlaminar shear which occurs when the resin matrix fractures and the individual plies separate.

COMPOSITE CONSTRUCTION

Composites are made up of laminates and sandwiches. Laminates are composite materials of two or more layers bonded together. Sandwiches are multiple layer structural materials which contain a low density core between thin faces of composite materials. Theoretically, there are as many different types of laminates as there are possible combinations of two or more materials.

Materials are divided into metals and nonmetals; nonmetals are divided into organic and inorganic. Accordingly, there are six possible combinations in which laminates can be produced: metal–metal, metal–organic, metal–inorganic, organic–organic, organic–inorganic, and inorganic–inorganic. If the laminates contain more than two layers, there are many more possibilities.

Sandwiches consist of a thick, low-density core, such as a honeycomb or foamed material, between thin faces of a high-strength and high-density material. In sandwich composites, a primary objective is high strength-to-weight ratio. The core separates and stabilizes the faces against buckling under torsion or bending, and provides a rigid and efficient structure.

PROPERTIES OF COMPOSITES

Composites can be made stronger than steel, lighter than aluminum, and stiffer than titanium. This is possible through the careful selection and use of high-strength fibers such as carbon/graphite, aramid, or boron, bound in a matrix of epoxy. In aircraft structures, a graphite–epoxy composite presents about the same strength and stiffness as aluminum; however, the composite weighs 45% less than aluminum; it has a superior fatigue resistance and a lower thermal conductivity; it is also noncorrosive and highly wear resistant.

However, corrosion could occur if graphite and aluminum are in direct contact in the presence of moisture. This reaction does not occur when graphite and titanium are in contact. The two properties most in demand when advanced composites are considered are tensile strength and Young's modulus. Aramid has a slightly higher tensile strength than carbon/graphite, but a much lower Young's modulus.

APPLICATIONS OVERVIEW

Advanced composites containing such exotic materials as carbon/graphite and aramid fibers in an organic resin matrix are being used currently, mainly by the aircraft industry. See Fig.

BORON COMPOSITE MATERIALS

F-14 HORIZONTAL STABILIZER
FIRST BORON COMPOSITE PRODUCTION UNIT

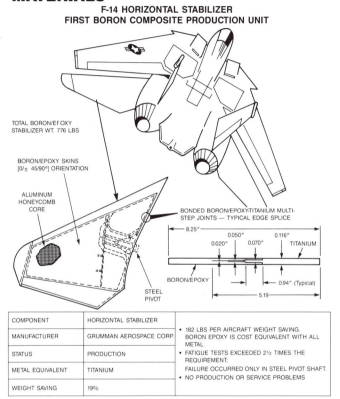

TOTAL BORON/EPOXY
STABILIZER WT. 776 LBS

BORON/EPOXY SKINS
[0/± 45/90°] ORIENTATION

ALUMINUM
HONEYCOMB
CORE

BONDED BORON/EPOXY-TITANIUM MULTI-
STEP JOINTS — TYPICAL EDGE SPLICE

8.25"
0.050"   0.116"
0.020"   0.070"   TITANIUM

BORON/EPOXY

0.94" (Typical)

5.19

STEEL
PIVOT

COMPONENT	HORIZONTAL STABILIZER	• 182 LBS PER AIRCRAFT WEIGHT SAVING. BORON EPOXY IS COST EQUIVALENT WITH ALL METAL. • FATIGUE TESTS EXCEEDED 2½ TIMES THE REQUIREMENT; FAILURE OCCURRED ONLY IN STEEL PIVOT SHAFT. • NO PRODUCTION OR SERVICE PROBLEMS
MANUFACTURER	GRUMMAN AEROSPACE CORP.	
STATUS	PRODUCTION	
METAL EQUIVALENT	TITANIUM	
WEIGHT SAVING	19%	

Fig. 24.2. Advanced composites in an aircraft structure. (Courtesy of Textron Inc.)

24.2. Nevertheless, other industries are using AC in applications ranging from automobiles, spacecraft, printed circuit boards, and sports equipment to prosthetic devices. These uses are for unusual situations because the cost is very high.

DATABASE LIMITATIONS

Thirty years of development in the advanced composite field have left us with a scarcity of data. We are still very limited in knowledge of how stress is carried and transferred in complex loads; the interface bond between the fiber and matrix is not fully understood; and impact resistance is another area in which we lack knowledge.

Metal will crack or at least dent when struck a blow. However, when a composite is struck and there is no external, visible damage, there may be a defect on the inside. Delamination may have started internally and it may propagate until destruction occurs. Until recently, only thermosetting resins were used as resin matrices because of their high temperature properties. Recently, several thermoplastics have become available for high-performance roles.

FABRICATION

Industry has discovered that the experience acquired in years of producing fiberglass would be somewhat successful in fabricating composites. Organic matrix composites are made primarily by molding in autoclaves, while metal matrix composites are formed by diffusion bonding.

There are four popular methods of producing continuous fiber composites with closely controlled properties: lamination, filament winding, pultrusion, and injection molding. The construction technique selected depends upon the shape, size, type of part, and the quantity to be manufactured.

LAMINATION

The laminate process starts with a prepreg material (partially cured composite with the fibers aligned parallel to each other). A pattern of the product's shape is cut out, and the prepreg material is stacked in layers in the desired geometry. The assembled layers are then cured under pressure and heat in an autoclave. Graphite/epoxy composites are cured at a temperature of 350° F and a pressure of 100 psi. The new high temperature composites such as bismaleimides are cured at 600° F.

FILAMENT WINDING

In the filament winding process, fibers or tapes are drawn through a resin bath and wound onto a rotating mandrel. This is a slow process, but the direction can be controlled and the diameter can be varied along its length. If working with tape, it is an endless strip whose width can vary from an inch to a yard. With both fiber and tape winding processes, the finished part is next cured in an autoclave and removed from the mandrel later.

In aerospace structures, which are strength-critical, carbon fibers are wound with epoxy-based resin systems. The other resin systems are limited to special applications. Filament winding is used to make cylindrical objects like missile cannisters, pressure bottles and tanks.

PULTRUSION

In composite technology, pultrusion is the equivalent of a metal extrusion. In the process, a continuous fiber bundle is pulled through a resin matrix bath and then through a heated die. The process is generally limited to constant cross sections such as tubes, channels, I beams, and flat bars.

INJECTION MOLDING OR RESIN TRANSFER MOLDING

This method fills the space between compression molding and hand manufacturing layup. In resin transfer molding (RTM), two-piece matched metal molds are used. RTM uses low injection pressures, which in turn allows the use of low cost tooling. The reinforcing material, either chopped or continuous strand, is draped in the cavities, the two-mold halves are clamped together, then the resin is pumped into the closed mold.

CAD/CAM

Research and development programs and, as a matter of fact, any mold design can be expedited by the use of CAD/CAM. Design changes can be tried out quickly this way with minimal risk. For instance, if you show the assembled mold on the screen, it will indicate interferences and restrictions, thus precluding re-machining and modifications. CAD/CAM has even been used to lay out equipment and complete floor plans for producing advanced composites.

MACHINING, CUTTING, AND JOINING

AC materials are generally unsuited for normal machining and fabrication techniques, so special methods must be used. Before doing any machining on AC materials, you must ensure that there will be no delamination, fraying, or cracking of cured composite edges. With modifications, standard machine tools can often be used. Spindle speeds and feeds should be selected depending on the type of laminate material and its thickness and the machine used. Whatever cutting tools are used, they must be sharp.

Uncured composite materials can be cut with shears, scissors, or carbide disc cutters. For cured composites, reciprocating knife cutters, ultrasonics, lasers, or abrasive waterjets may be used. Lasers work fine on cured composites, but may burn uncured material; abrasive waterjet cutting may present moisture problems; knife cutters can clog. The safest procedure is to question the suppliers.

If the structural composite is made with a thermosetting resin, it cannot be joined by welding as if the resin were thermoplastic. Joining would have to be by adhesives or mechanical means. Sometimes both are used.

Defects in composites, such as cracks, are sometimes easily detected; others can be difficult to find. Voids, missing layers, delaminations, inclusions, and improper layup present problems. Most metals fail by fatigue, but composites break under load, and this could lead to catastrophic failure. Since composite structures vary in point-to-point comparisons, selecting samples for destructive testing would have no value. Also, since fiber reinforcements and resins can appear the same in an X-ray, that medium is ineffective. Therefore, we are left with only a select few methods for effective quality control.

Conventional radiography will provide good resolution when the attenuation characteristics of fiber and resin are quite different, as in boron epoxy composites. However, in aramid epoxy or graphite epoxy composites, where differences are small, defect determination is difficult.

Ultrasonics is useful in detecting skin and bonding problems. Liquid coupled ultrasonics is currently the most widely used inspection method. Thermography, optical holography, and eddy current testing are also used in special cases. One precaution should be taken when processing fibers or yarns: avoid inhalation of airborne particles. Dust from this environment may cause respiratory difficulties. Dispose of dust and remnants at a suitable landfill because this material is difficult to burn.

Temperature control is of utmost importance in producing CVD SiC fiber, as it is in any vapor deposition process. Avco, for instance, uses a temperature of about 2370° F. Temperatures above this cause rapid deposition and subsequent grain growth which lowers strength. Temperatures below the optimum cause high internal stresses in the fiber, resulting in a degradation of metal matrix composite properties if machining is done transverse to the fibers.

Various grades of fibers are produced, which are based on the standard BSiC deposition process described earlier, where a crystalline structure is grown onto a carbon substrate.

Substrate quality is another important consideration in SiC fiber quality. The carbon monofilament substrate, which is melt-spun from coal tar pitch, has a very smooth surface with occasional anomalies. These anomalies, if severe, can result in localized irregular deposition of PG and SiC (this is a stress raising area). The carbon monofilament spinning process is closely controlled to minimize production of low-strength fibers.

The PG flaw results from insufficient control of the CVD process. This flaw is caused by irregularities in the PG deposition. There are two reasons for PG flaws: an anomaly in the carbon substrate surface, and mechanical damage to the PG layer prior to the SiC deposition. The surface of Avco's SiC fibers is carbon rich, which protects the fiber from surface damage. Surface flaws can be identified by an optical examination. All flaws are minimized by careful handling and close adherence to fabrication instructions. Typical mechanical properties of the Avco CVD SiC fiber are: 575 ksi and an elastic modulus of 60 msi.

FIBER VARIATIONS

It is important to tailor the surface region of the SiC fibers to the matrix. There is a difference in the surface composition of three fibers. SCS-2 has a carbon-rich coating which increases in silicon content as the outer surface is approached; this fiber commonly reinforces aluminum. SCS-6 is primarily used to reinforce titanium; it has an even thicker carbon-rich coating in which the silicon content also increases as the outer surface is approached. SCS-8 was developed to provide better mechanical properties in aluminum composites than SCS-2.

COST FACTORS

SiC is potentially less costly than boron for three reasons.

1. The carbon substrate used for SiC is less expensive than the tungsten used for boron.
2. Raw materials for SiC (chlorosilanes) are less expensive than boron trichloride, the raw material for boron.
3. Deposition rates for SiC are higher than those for boron.

In 1988, continuous silicon carbide fiber cost $2,500/lb with 30,000 feet of 5.6 mil fiber weighing 1 lb. It was estimated that the fiber could cost as little as $100/lb if full-scale production reaches 40,000 lb/year. Five contractors on projects funded by NASA have been furnished samples of silicon carbide reinforced titanium. It is hoped that airplane manufacturers will accelerate the use of composites and bring the price down to the point where other industries will make use of them.

Borsic and boron fibers have been evaluated for use in aluminum alloys; and unless complex, high-pressure low-temperature, diffusion bonding procedures are followed, degradation of fiber strength will result. Similarly, with titanium, unless fabrication times are shortened, fiber/matrix interactions will produce brittle, intermetallic compounds that severely reduce composite strength.

The SCS grade of fiber, in contrast, has surfaces that bond readily to various metals without destructive reactions. This results in the ability to consolidate aluminum composites with investment castings and low pressure molding. Similarly, for titanium composites, the SCS-6 filament can withstand long exposure at diffusion bonding temperatures without fiber degradation. Accordingly, complex shapes with these composite reinforcements can be fabricated by the superplastic forming/diffusion bonding and hot isostatic pressing process.

COMPOSITE PREFORMS AND FABRICS

An old system consisting of a single layer of fibers that are spaced side by side, held together by a resin binder and supported by a metal foil, is called green tape. This layer is a pre-preg which can be laid up in a mold in required orientations to fabricate laminates. The laminate processing cycle is then controlled to remove the resin by vacuum as volatilization occurs. The method generally used is to wind the fibers onto a foil-covered rotating drum, overspraying the fibers with the resin, and finally cutting the layer from the drum to provide a flat sheet of prepreg.

Plasma sprayed aluminum tape is a more advanced prepreg than green tape. It replaces the resin binder with a plasma sprayed matrix of aluminum. There are two advantages: a lack of possible contamination for resin residue, and faster processing time because hold time to ensure volatilization and removal of the resin binder is not required.

Woven fabric is a uniweave system in which the large diameter SiC monofilaments are held straight and parallel, collimated at 100–140 filaments per inch, and held together by a crossweave of low density yarn or metallic ribbon. There are now two types of looms which can be modified to produce the uniweave fabric: the Rapier-type loom and a shuttle-type loom.

PROCESSING METHODS

Investment casting is a manufacturing technique used for many years as an inexpensive method of producing complex aluminum shapes. The aerospace industry has hesitated to use the process aggressively because, rightly or wrongly, the process was suspected of lacking consistent strength. However, since the material is now fiber dependent, interest in investment casting has been revived.

At present, the investment casting mold is opened, and the fibers positioned inside (after removal of the wax form, which is the shape of the desired casting). Hot molding is a term used by Avco to describe a low pressure, hot pressing process that is used to make SiC-aluminum parts cost-effectively. So we now have two new techniques for using fiber reinforcements.

MAKING RELIABLE COMPOSITE JOINTS

Joining composite materials presents special problems. The usual methods of joining are mechanical fasteners, adhesives, or both together; it depends on the composite and the application. Where failure could be catastrophic, as in aircraft, composites are generally joined by a combination of adhesives and mechanical fasteners. In other applications, adhesives are most common. Many engineers and designers, as well as users, distrust adhesives, which should be the first choice for joining composites. Actually, where large forces are encountered, adhesives spread out the stress, whereas fasteners tend to concentrate the stress.

Many factors should be considered when contemplating fasteners for composites.

1. When a spread of temperature is expected, the difference in coefficient of expansion could be significant.
2. If there is a possibility of delamination either by load stresses or the physical act of drilling.
3. When moisture invasion around or under the fasteners might cause galvanic corrosion.
4. If sealing may be necessary.

Potential changes in clamping forces should be determined at the design stage. Metal fasteners expand and contract with temperature changes, so it would be wise to consider this effect on high strength joints. In many cases, this effect demands the use of adhesives.

You should perform as little machining as possible on composites. Drilling or milling, for instance, could cause delamination, fiber breakout, or resin erosion. Each composite has its own machining problems. Carbon fiber materials require carbide drills and cutting tools; they are prone to delamination and fiber breakouts and create much dust. Aramid does not suffer from these problems, except for delaminations. Instead, aramids can melt when drilled, and the edges may fray.

Fasteners for composites should have large heads so that the loads can be distributed over as large a surface as possible. Clearance holes for fasteners should closely fit the hardware to reduce the tendency of fretting. Interference fits, however, may cause delamination. Special sleeve fasteners are available to provide interference fits while limiting chances for damage.

Cutting fibers exposes ends to moisture absorption which weakens the material and adds undesirable weight. Sealants could be used to prevent water absorption providing there is no necessity for electrical continuity. Since most composites are not conductive anyway, this should not present a problem. As a matter of fact, nonconductivity can present a problem in cases where electrical conductivity is desired.

Aluminum fasteners should never be used to fasten carbon fiber composites because galvanic corrosion is caused by a chemical reaction between the fibers and the fasteners; though coating the fasteners will prevent the reaction. Or, if preferred, aluminum fasteners can be replaced with titanium or stainless fasteners.

Three types of adhesives are commonly used to bond composites: epoxies, acrylics, and urethanes. Generally, epoxies are used with epoxy-based composites because they have similar flow and expansion characteristics. When large parts are being adhesive bonded, room temperature curing is an asset. Epoxies should not be used to join flexible composites, unless a flexibilizing agent is added to the epoxy. Polyurethane adhesives make flexible joints; however, they are sensitive to moisture and require complex dispensing equipment. Acrylics are rigid and cure quickly at room temperature; however, they have a bad odor, poor impact resistance at low temperature, and are flammable.

Recommended composite surface preparation is to solvent wipe and abrade gently. When bonding composites to metal, the metal should be prepared the usual way. Clean it by abrading, sanding, or blasting, and apply the adhesive soon after, before the surface oxidizes or gets contaminated.

25

ULTRASONIC TECHNOLOGY

Ultrasonic energy can be used for production in a number of different ways. The most common are machining, welding, hardware insertion, staking, and spot welding.

ULTRASONIC MACHINING

Ultrasonic machining (impact grinding) provides the capability of machining hard, brittle materials in jobs not feasibly done with more conventional means. The process is nonchemical, nonthermal, and nonelectrical, so that the parts being processed are not affected metallurgically. Slots, irregular shapes, blind cavities, and through cavities can all be machined by this method.

The actual method involves the use of an abrasive slurry, such as silicon or boron carbide, which flows between the workpiece and a tool. The tool moves vertically only a few thousandths of an inch while vibrating about 20,000 times per second. The flow of abrasives under and around the tool permits particles to strike the workpiece, chipping off tiny pieces of material until a counterpart of the tool is imprinted in the workpiece. See Fig. 25.1.

Common materials machined ultrasonically are glass, quartz, fiber optics, zirconia, germanium, ferrite, alumina, and other ceramics and hardened steels. It is not difficult to drill holes in these hard and brittle materials and hold 0.005″ tolerance on diameters. When the ultrasonic machine is configured to appear like a Bridgeport milling machine, it can maintain close tolerances on hole location and edge machining.

ULTRASONIC WELDING

Ultrasonic welding is used to join plastic parts. When two plastic parts are mechanically vibrated together at the ultrasonic

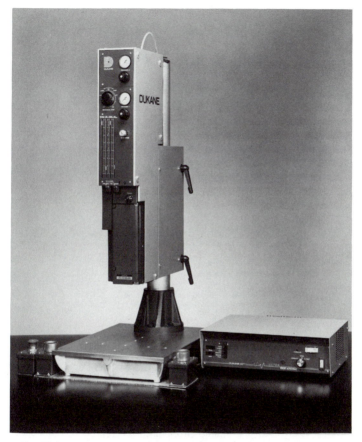

Fig. 25.1. Heavy duty thruster/press and 1000 watt generator.
(Courtesy of Dukane Co.)

frequency, friction generates localized heat that causes the materials to melt, flow, and fuse. (See Fig. 25.2 for weldability of
various thermoplastics.) The equipment required for this process consists of a power supply (generator), transducer, booster,
horn, and handgun. This equipment can be integrated by

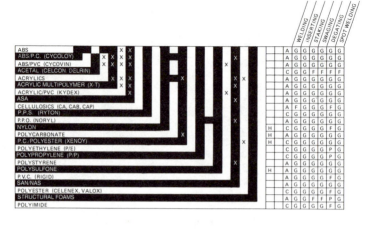

WELDABILITY – READ ACROSS   COMPATIBILITY – READ ACROSS AND UP

■ = GOOD COMPATIBILITY   X = COMPATIBLE AT TIMES BASED ON MATERIAL COMPOSITION

H = HYGROSCOPIC, SHOULD BE DRY BEFORE WELDING   A = AMORPHOUS RESINS   C = CRYSTALLINE RESIN

WELD CHARACTERISTICS: G = GOOD   F = FAIR   P = POOR

Fig. 25.2. Weldability chart for thermoplastics. (Courtesy of Dukane Co.)

sophisticated automation into a system of repeatable production.

An ultrasonic generator accepts 60 cycle current and delivers 20,000–40,000 Hz to a piezoelectric transducer. Piezoceramic wafers convert the electrical energy to mechanical motion. This motion is then imparted to acoustically resonant tools called horns. When the horn is placed on the workpiece, it passes vibratory energy to the joint, producing a controlled melt at the point of least resistance. The weld cycle lasts about one second. For maximum strength, some type of shear joint should be used.

SHEAR JOINTS

The basic shear joint is shown in Fig. 25.3, before, during, and after welding. The Du Pont Company recommends it in its book on plastics. Figs. 25.3–25.5 show three variations of the basic shear joint.

Shear joint

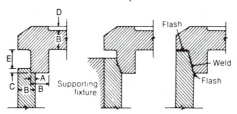

Fig. 25.3. Shear joint before, during, and after welding.

Modifications for large parts
(Shear joint)

Fig. 25.4. Large parts may require modification before welding.

Flash traps
(Shear joint)

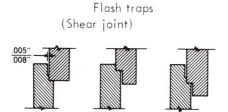

Fig. 25.5. Examples of flash traps.

In forming the shear joint, it is best to limit contact to a small area, which could be a recess or a step in one of the parts. First, the contact surfaces melt; then, as the parts telescope together, they continue to melt along the vertical walls. The smearing action of the two melt surfaces precludes leaks and voids, making this the best joint for leak-free seals. It is also the fastest joint to make and the one requiring the lowest energy. Heat generated at the joint is retained until vibrations cease because, during telescoping and smearing, the melted plastic is not exposed to air which would cool it.

Weld strength is a function of weld time and part design which create the depth of telescoped section. By designing the depth of telescoping walls at 1.25–1.5 times the wall thickness, you can ensure a weld stronger than the welded material.

There are several significant features to this type of design which should be considered. The top part should be shallow and as close to a lid as possible. You should have a fixture to hold the bottom part to keep it from expanding during welding. If possible, the fixture should also prevent the upper part from slipping out of position. If the contact area is too small, the design shown in Fig. 25.4 should be considered. The design should make provision for flash when it cannot be tolerated. Fig. 25.5 shows where displaced molten material could be directed.

The statement that the weld is stronger than the rest of the part can be true only if the instructions given above are followed. Welding provides safer working conditions by eliminating the need for toxic solvents and messy adhesives. Welding creates consistent, strong, integral bonds, and it lowers rejection rates. It saves part and labor costs by eliminating hardware.

BUTT WELDS

The common butt joint (shown in Fig. 25.6), which includes a weld concentrator, is very popular. Upon application of ultrasonic energy, material from the concentrator provides a solid weld as it disperses throughout the weld area. The base of the concentrator should be about 20% of the overall width, but not over 0.025" high.

The Du Pont Company has issued a notice referring to Delrin, Zytel, and Lucite. They suggest avoiding the butt weld for for these materials because they crystallize before sufficient heat is generated to complete the weld.

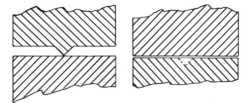

Fig. 25.6. A butt weld.

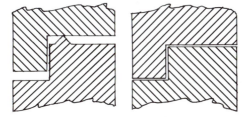

Fig. 25.7. A step weld.

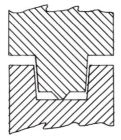

Fig. 25.8. A tongue-and-groove weld.

STEP WELDS

Using the step design will strengthen the weld because the melted plastic from the concentrator moves into the slip fit area. This weld generally provides a clean appearance. See Fig. 25.7.

TONGUE-AND-GROOVE WELDS

This is a combination of both the butt and step welds. It presents a good appearance with excellent strength. See Fig. 25.8.

WELD VARIATIONS

You can adapt many variations of the butt, step, and tongue-and-groove welds. Modifications are sometimes necessary in order to obtain a special weld.

ULTRASONIC HARDWARE INSERTION

Very often, it is necessary to get metal hardware into plastic parts; i.e., usually small metal parts, like threaded inserts or pins, must be embedded. This is done by premolding a hole in the proper position in the plastic piece which is slightly smaller than the component to be embedded. This hole guides the metal insert and creates an interference fit between the insert and the plastic piece. As soon as the melt occurs at the interface, the melted plastic flows into the knurls and undercuts of the metal insert to lock in the encapsulated metal part.

There are advantages to using ultrasonics for insertions. First, the energy is generally activated for no more than a second. Second, very little stress is created in the surrounding area, thus eliminating cracking while maintaining dimensions. Third, the inserted component normally exhibits high torque and pull-out strength. Finally, the process parameters can be consistently controlled.

ULTRASONIC STAKING

This is an assembly procedure used to join dissimilar materials: usually metal to plastic or plastic to plastic. Generally, the plastic part will have protrusions or posts which extend up through holes in the metal piece. A specially contoured horn contacts the post and melts it so that it reforms to establish a locking head over the metal part. This is also the scenario if the two parts are plastic or if a plastic rivet is used.

In any process involving localized heating by ultrasonics, the designer must control where and how fast the temperature rise will occur. Geometry plays a vital role in determining the location of high strain which creates a desirable heating zone (see

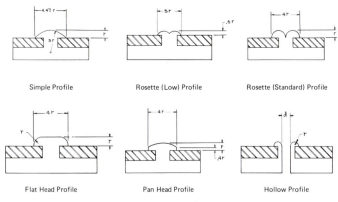

Simple Profile   Rosette (Low) Profile   Rosette (Standard) Profile

Flat Head Profile   Pan Head Profile   Hollow Profile

Fig. 25.9. Staking configurations.

Fig. 25.9). Therefore, an energy director should be employed when using the ultrasonic staking technique.

ULTRASONIC SPOT WELDING

This type of welding is recommended when two thermoplastic parts are to be joined at localized places, thus eliminating the need for more complex joints.

MISCELLANEOUS

There is a good number of highly qualified ultrasonic machine manufacturers in this country making equipment similar to that described above, so you should have no trouble obtaining competent advice. We have noted several different manufacturers during our discussion, any one of whom could perform to your requirements.

ANCILLARY EQUIPMENT

There is a number of small ancillary units marketed by various companies producing ultrasonic equipment. For example, a general-purpose probe system has been designed to be used for small fragile parts; its power source is 120 V, 50/60 Hz; the output power is 350 or 700 W with an output frequency of 40 Hz; and it can be adapted to continuous seam welds. Several other units are described below.

PARTS HANDLING SYSTEM — ROTARY

Dukane makes a rotary parts handling system which is designed for economical manual or semi-automatic operation with free-standing base, welder, six-station rotary table, dial plate, and safety guard. This system can perform spot welding and any other common welding, staking, inserting, and swaging for medium- or heavy-duty service.

ULTRA CUTTER

The Ultra Cutter is designed to cut the new aerospace composite materials such as graphite fibers, aluminum honeycombs, and kevlar. A 1″ blade vibrates at 40,000 cycles per second which cuts the materials with minimum force. At the same time, the cut edges of fabric materials are fused to prevent unraveling.

MICROCOMPUTER

This unit helps automate the use of ultrasonics. It has a modular design which permits simple field installation of interface features. It is a real-time, multitask, operating system which simultaneously controls and monitors multiple process parameters. It provides input to a printer for permanent documentation and to a screen display for viewing.

HORN ANALYZER

This can be very useful in the manufacture of trial and production horns of both 20 and 40 kHz frequencies. It also serves as a valuable trouble-shooting tool for existing ultrasonic equipment.

In conclusion, ultrasonics is another field in the arena of sophisticated processes which is destined for increased popularity in the near future.

26

HOT ISOSTATIC PRESSING

BACKGROUND

Hot isostatic pressing (HIP) is a process of manufacturing which has acquired commercial production status relatively recently. During the 1950's, it was only a laboratory procedure used to make synthesized industrial diamonds. Gradually, applications were found which required process improvements. Those improvements added up until now there are several applications where this process is outstanding. These applications include the compaction of powders—both metal and ceramic—and, more recently, the densification of castings and the elimination of microshrinkage porosity. Howmet Inc. of Whitehall, MI, is the first foundry in the country to establish its own production facility to HIP castings. It has launched extensive development research to establish Howmet as a leader in casting and HIPing titanium and superalloy products especially for the aerospace industry. See Fig. 26.1.

PROCESS

Hot isostatic pressing is the simultaneous application of heat and pressure created by heating inert gas in a container (see Fig. 26.2). This use of gas pressure provides uniform force all over the parts in the container; and it is done without using expensive dies. HIP was originally developed to join metal parts by diffusion bonding, and this bonding is still a vital attribute of the process.

Currently, the process is used to create denser parts and materials. Castings and powder (metal and ceramic) parts all contain internal voids which often seriously affect mechanical properties; the HIP process will close voids and then heal them

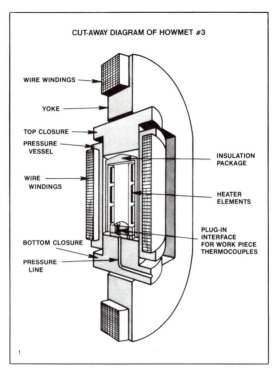

CUT-AWAY DIAGRAM OF HOWMET #3

WIRE WINDINGS

YOKE

TOP CLOSURE

PRESSURE VESSEL

INSULATION PACKAGE

WIRE WINDINGS

HEATER ELEMENTS

PLUG-IN INTERFACE FOR WORK PIECE THERMOCOUPLES

BOTTOM CLOSURE

PRESSURE LINE

1

HOWMET HIP CAPABILITIES

	# 1	# 2	# 3
WORK DIAMETER	20"	10"	41"
WORK HEIGHT	69"	20"	96"
TEMPERATURE CAPABILITY	2250F	2640F 3180F*	2300F
PRESSURE CAPABILITY	30KSI	15KSI	15KSI
HEATING ELEMENT	Mo	GRAPHITE	Mo
WORKLOAD T/C CAPABILITY	16	8	25

ON-LINE GAS CHROMATOGRAPH TO ASSURE 99.99% PURE ARGON.
25F (14C) TEMPERATURE CONTROL CAPABILITY.

*WORK ZONE RESTRICTED TO 4□ x 8 HIGH

2

Fig. 26.1. Howmet HIP furnace. (Courtesy of Howmet Corp.)

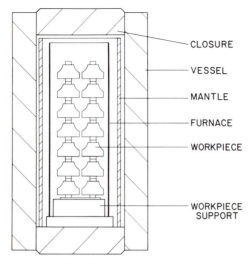

Fig. 26.2. Schematic of HIP vessel.

by diffusion bonding. The mechanical property most improved by this process is fatigue strength. The voids in processed parts are similar to high stress notches which cause premature fatigue failures.

Encapsulation of parts is often a necessary first step. A part may be made by mechanically pressing powder or by cold isostatic pressing (CIP); or it might be made by the injection molding process (see Chapter 1). No matter which process is used to prepare the "green" part, internal voids are created which might be connected to surface porosity. If this were the case, it would lead to internal contamination by gases during pressurization, and this cannot be tolerated.

Consequently, many parts must be securely sealed against gas penetration before pressurization; this includes parts which have already been sintered. Sheet metal encapsulation is a very popular method of protection. Sometimes encapsulation is performed by heavy walled metal molds or ceramic molds. Glass is also commonly used to encapsulate parts to be HIPed.

The encapsulation technique selected depends on several

features, not the least of which is cost. The utilization of work-load space in the furnace is important. The more parts placed in the furnace in the same load, the less expensive the cost per part. There are three common encapsulation methods described below.

The first method is to place the "green" body in an oversized capsule of silica or boron silicate glass. (One precaution: care must be taken to avoid contamination. This glass can react with some materials.) The capsule is evacuated at an elevated temperature before being heat sealed; then the heated glass conforms to the shape within. During the HIP process, the heat and pressure compress the part within the glass without allowing gas to penetrate surface porosity of the part.

The second method is to immerse the porous parts in powdered glass which has a low softening temperature. This method avoids the complications of method one, but it has a complication of its own: glass particles might penetrate the very pores you are trying to protect. Accordingly, a special material should be put over the entire surface of the parts being prepared for the HIP cycle. This step prevents glass from entering surface pores.

The third method is to carry the parts through a sintering step and then HIP them. Normally, parts which have been sintered have closed surfaces and do not require encapsulation prior to HIP; that permits you to load more parts per HIP cycle. The cost per HIP cycle depends on the size of the container. A large container might cost $4000 per cycle; a medium-sized container might cost $2000 per cycle; and a small one could cost $1000. Naturally, one should compute the quantity of parts which could be HIPed in each size of container and then proceed with the least expensive.

There is another HIP process, called high pressure reaction sintering. This uses high pressure nitrogen gas instead of argon, which is the most commonly used inert gas. Nitrogen sometimes provides better results than argon. However, when processing nitride ceramics, nitrogen gas could react with carbon materials and form a poisonous gas. Suitable precautions must be taken to prevent such an occurrence.

The pressure and heat of hot gases during the HIP process is what provides the uniform consolidation of porosities, even those appearing in weld joints. These pressures and temperatures can be adjusted during the HIP cycle. The cycle may begin with a heating step to initiate the encapsulation; then it continues in a prepared sequence which makes the cycle ideal for computer control.

As stated previously, the inert gas used in this process is most often argon. Each different material processed requires an adjustment of pressure, temperature, and time to achieve optimum results. The part preparation process requires its own specific adjustments. For instance, a stainless steel casting has a different set of adjustments than the same material in powder form. The HIP settings for a sintered part are different than the settings of the same part which is dry pressed or cold isostatic pressed.

COLD ISOSTATIC PRESSING (CIP)

This process uses high pressure, through hydraulics, to densify powders in rubber molds, to approximately 95% of theoretical density. Then the parts can be either sintered or put through the HIP process. If mechanical properties are significant, the HIP process will raise density to nearly 100%.

In the CIP process, no heat is used, but the rubber mold is submerged in water so the pressure is uniformly applied all over the mold just like the pressure in HIP. Some vendors supply both CIP and HIP services. They should be consulted about the sequence of steps your product should have before the part drawings are cast in concrete.

THE ROLE OF HIP IN INDUSTRY

HIP has become exceedingly significant to modern industrial technology. It is basically a batch process, but it has increased its output per batch cycle from pounds to tons. Its required use is noted on drawings and purchase specifications. This increased popularity is, of course, due to its unusual results (increased mechanical properties), which in turn are due to innovations in the design and construction of HIP equipment.

A wide range of applications for HIP has become apparent due to extensive experimentation by a few dedicated vendors and scientific communities (like Battelle). The most important applications of HIP in current use are:

Densification of internal flaws in many materials, including castings, powdered metal and ceramic parts, and welments;

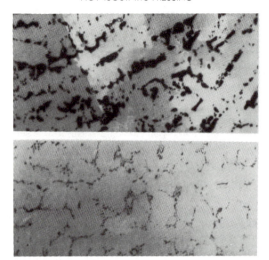

Fig. 26.3. Densification of castings by HIP (100X). Before HIP (top) and after HIP (bottom).

The diffusion bonding of similar and dissimilar materials and the healing of internal porosity; and

Powder compaction production of both metal and non-metallic parts.

HIP equipment basically consists of an electric furnace within a high pressure vessel. This equipment can perform functions which permit improvements in other processes. A few years ago, the author had parts made by metal injection molding. It was a risky thing to do because the same parts, machined from bar stock, had borderline strength. The parts, made by MIM and then put through an HIP process, performed successfully and saved considerable money.

HIP AS RELATED TO CASTINGS

HIP equipment is permitting changes to take place in the casting industry. Investment casting was a process much desired by the aeronautic industry and others because its products could be used with a minimum of machining. However, users insisted upon better and more consistent mechanical properties. With HIP, these properties do improve, especially fatigue strength, and their values are more consistent. See Table 26.1.

For example, the ratio of part weight to gating weight used to run as high as 1:6; now this ratio is down to 1:3. When making large castings of titanium or a superalloy, the weight saved adds up to a large sum.

Also, when high temperature creep strength is a job requirement, casting is an excellent process to select. Although there still are problems to be eliminated, the combination of casting plus HIP can solve creep strength problems. The diffusion bonding of HIP provides metal integrity. As a consequence, HIP is now written into many bills of material.

Now, designers of castings actually anticipate the advantages of HIP which allow them to use less gating and make a variety of other changes in the casting procedure. HIP will permit castings to replace more expensive forgings in some places.

In permanent mold casting, the skin chills quickly and should not be porous; this will permit the HIP process to be used without requiring any encapsulation. As things stand presently, encapsulation adds enough cost to the process to price it out of competition in too many cases. Research is being done to bring its cost down. Sources are working on highly viscous glasses and ceramics for encapsulation.

Internal shrinkage and porosity are the major reasons for rejection of stainless steel castings. The internal defects are detected by radiography, while dye penetrant testing detects surface imperfections.

Castings have an inherent problem with porosity which the industry has lived with for decades. Since HIP became available as a tool to heal internal flaws, much attention is being paid surface welding as a means of sealing to avoid conventional encapsulation. This overlay is routinely done to titanium, stainless steel, and other types of castings.

One of the steps which create expense in the HIP process is cooling. Research is being done to accelerate the cool down part of the HIP cycle. Besides lowering expense, it may also allow heat treatment of castings to take place at the same time, during the cooling cycle.

Industrial Materials Technology Inc. of Andover, MA, is another company providing HIP service to industry. The following is a typical sequence of events at their plant.

Table 26.1. HIP Increases the Strength of Castings

Alloy	Condition	Fatigue Life, cycles @ 20,000 psi	Yield Strength, psi	Ultimate Tensile Strength, psi	Elongation at Rupture, %
Sand Mold Cast C355	As-Cast + T6	147,500	32,400	34,200	0.8
	HIP + T6[a]	2,905,000	35,200	40,000	1.8
Permanent Mold Cast A356	As-Cast + T61	452,000	28,900	38,500	7.5
	HIP + T61[a]	4,650,000	30,500	42,200	11.1
Permanent Mold Cast 142	As-Cast + T4	709,000	29,100	32,600	—
	HIP + T4[a]	10,346,000	29,600	36,000	—

[a]15,000 psi
Source: Alcoa

1. A fixture is loaded with as many castings as can fit within the container. The fixture and/or support structure must prevent distortion to the castings.

2. The fully instrumented load is transferred into the pressure vessel, and the insulation system is positioned.

3. The cold system is closed, evacuated, flushed with argon, and evacuated again.

4. Normally, the container is pressurized to some initial pressure so that the desired operating temperature and pressure are reached at about the same time.

5. Dwell time at peak pressure and temperature will vary from 2 to 4 hours. When the power is turned off, the system cools by itself. The system is so thermally tight that this part of the cycle is the longest.

The choice of settings (dwell time, temperature, and pressure) is made experimentally. The first estimates are made on the basis of experience and engineering data. The control of temperature is more critical than the other two parameters. Gas purity is also critical. New gas for each run would be too expensive, so it is generally recycled. For this reason, on-line analytical tools are constantly monitored to detect contamination of the inert gas.

It is hoped that research and development will solve the problem of surface connected internal flaws and the long cooling cycle. Solutions to these problems would make HIP an economical step to be used more often because the difference between acceptability and rejection of a casting is generally internal flaws. Radiographic examination locates the flaws and HIP is used to salvage the castings. Generally, it is only the expensive castings which are salvaged. But if costs decline, more borderline castings would use HIP for repair.

The major attraction in using HIP is to improve mechanical properties. Since fatigue strength, yield, and tensile strength are significantly improved, designers can work with higher values.

CASE HISTORIES

Each vendor you speak to can recount tales of successes which will interest you. Perhaps one of the following case histories will parallel your problem.

Case 1: Rene 120 high pressure turbine blades had been plagued with poor castability. Dovetail grinding exposed shrink porosity causing a high reject rate and a high cost. HIP reduced the rejects from 28% to 4%.

Case 2: Prior to 1986, the largest castings which could be densified were 42″ in diameter; this was too small for some jet engine castings. As a result, many small parts had to be assembled at high cost. Industrial Materials Technology Inc. installed a unit 60″ in diameter × 120″ long; now castings up to 60″ in diameter can be HIPed. This act precludes the necessity of making and assembling many individual parts, and has saved General Electric and Pratt Whitney considerable money.

Case 3: Hydroplane propellers are subject to severe stress in service, and sometimes fail due to impact or fatigue. Mercury Marine now uses HIP extensively on their propellers because their test results with HIP were so successful.

QUALITY CONTROL

Problem areas in the HIP process have been identified, and procedures have been established to monitor them. There are five main areas of HIP quality control.

1. One of the most significant variables in HIP processing has always been gas purity. Small quantities of oxygen, nitrogen, methane, carbon monoxide, and hydrogen, even in trace concentrations, can cause undesirable reactions. The Howmet Corp., for instance, has established limits (PPM) that control total impurities to a maximum of 100 PPM. These purity levels are monitored by means of an on-line gas chromatograph.

2. Temperature tolerance ($\pm 25°F$) is critical in HIP applications. Thermocouples and test cubes help maintain the required accuracy. The cubes are examined metallographically to confirm the thermocouple readings.

3. Pressure is maintained within desired limits by referencing the output of a pressure transducer.

4. Metallography is checked by including at least two bars of casting material in each run. The bars are tested in many ways to verify the quality of the castings.

5. Mechanical testing is employed to establish optimum production cycles, and then it is continued on a regular basis to meet customer requirements.

There are many emerging applications which could become commercial successes if the cost of HIP could be reduced. There have been case histories which indicate that the costs of part rejection, repairs, and reprocessing have outweighed the cost of HIP. Industry is interested only in the bottom line, so as soon as it is economical to HIP, the process will see greater use.

27

LAPPING, POLISHING, HONING

Three forms of surface finishing—lapping, polishing, and honing—are tried and true. They also used to be very expensive, and for this reason designers avoided them. But very serious problems can be solved by these processes. An example of such a problem solved by lapping involved a part, about $3'' \times 1'' \times 0.125''$ thick, full of holes, that was produced by fine blanking and then finished by double-disc grinding. The end product was consistently just enough out-of-flat to require a straightening operation. That part is now being lapped in a twin-wheel lapping machine. The bottom line is that the new process is producing flat pieces and saving money. Lapping, polishing, and honing in modern machines are so fast that these processes are being reintroduced to American industry.

DEFINITIONS

LAPPING

This is a low velocity abrasive machining process which uses loose rolling abrasives in a liquid or paste vehicle. A slight pressure and a relative motion are applied by the lap to the surface to be machined. The workpieces are free to align themselves against the lap. The process permits an accurate, nonreflective surface finish.

FLAT HONING

This is a high precision, low velocity abrasive machining process using bonded abrasives and flat honing plates (wheels). Pressure and motion are applied between the honing plates and the workpieces which are free to align themselves. Honing oil is used to lubricate, clean, and cool the parts and equipment. Modern machines work so quickly and to such close tolerances that they come equipped with refrigeration to help maintain a uniform temperature. There is no loose abrasive used in honing, rather there is fixed abrasive in the form of a wheel.

FLAT PRECISION POLISHING

Hard workpieces such as ferrites, ceramics, and tungsten carbides are polished in a manner different from soft workpieces like aluminum, silica, quartz, and silicon. A high precision, low velocity abrasive machining process which utilizes abrasive embedded in a hard polishing lap is used for the hard materials. Pressure and motion are applied between the polishing lap and the workpieces. For the soft materials, a relatively soft pad adhered to the flat polishing wheel loads up with soft abrasive and improves the finish and flatness of the workpieces. Controlled lubrication is used in the polishing, which results in a reflective surface.

EDGE HONING

In this process, there is a resilient pad bonded to the lower wheel. The pad is saturated with abrasives and pressure is applied by the upper wheel. The workpieces are partially submerged in the (hard) resilient pad; this causes the part edges to be honed or deburred in a clean and dry process. Note that in most new finishing machines, two wheels are used. They are required to achieve parallelism, and they generally turn in opposite directions. The pressure used in most cases is about 2–4 psi. Most of these machines are computer controlled for automatic operation.

INTERNAL HONING

This is a low velocity abrasive machining process utilizing bonded abrasive hones to do the cutting. Either the workpiece or the hone is free to align itself. Pressure and motion are applied between the workpiece and the hone. At the present time, the Sunnen Corp. sells a fixture which allows outside diameters also to be finished on the standard Sunnen (internal) honing machine.

WORKHOLDERS

Workholders are used to guide workpieces between the revolving hones, laps, or polishing wheels. They guide each part across the full face of the wheels in order to maintain uniform wheel wear. The parts should be able to turn freely in the workholder and align themselves between the wheels. Typically, workholders are sprocket carriers, driven in an epicyclic manner by a rotating inner ring of pins around a nonrotating outer pin ring. The speed and direction of the inner pin ring can be adjusted for optimum performance.

TRUEING AND DRESSING

Lapping and honing wheels are trued occasionally by a device holding industrial diamonds. They can also be dressed between trueings by having the top wheel rub lightly against the bottom one. Usually this is done with the top wheel slightly offset.

LAPPING

LAPPING TO CLOSE TOLERANCES

When the tolerance is less than 0.00005", time plays a very significant role. The operator handling the machine soon develops an understanding of the time/stock removal ratio. The cycle is stopped frequently to turn the parts end for end and to allow the operator to measure a part and record the size. The first cycle always removes the most stock since the previously machined finish has many peaks and valleys and the peaks lap off easily; also, the lapping compound is sharpest when the process first starts.

The abrasive grains begin to break down and cut more slowly, so each 30 second cycle removes less stock. The operator is able to machine to very close tolerances because of the reduction in stock removal rate. For example, the first 30 seconds might remove 0.00005"; the second cycle would remove 0.00003"; the third 0.000025", etc. The final cycle could remove only 0.000003"–0.000005" of stock.

STOCK REMOVAL

Normally, the amount of stock removed is between 0.00015" and 0.0005". If a decent grinding job has been done prior to the lapping, the lapping should remove imperfections in the part. If a finish of less than 1 microinch is desired, fine abrasive should be used, and the stock removed would be about 0.00015"; if the required finish is 2 microinches, then the stock removed would be about 0.0003".

LAPPING DEDUCTIONS

The material removal rate is proportional to the following influences: lapping speed, wheel pressure, grit type and size, type of lapping liquid, and proportion of grit per unit of liquid.

The volume of stock removed naturally increases as the speed of the wheel increases. Since the depth of penetration of the grit does not change, the surface finish is normally unaffected by the speed of lapping. However, it can be easily seen that the depth of penetration does change with increased pressure.

LAPPING ACCURACY AND FINISH

It is possible to attain accuracy and straightness to 5 millionths of an inch by lapping. The lap flatness and the operational procedure are the two factors which determine the machine capability. With the proper abrasives and lubrication selection, finishes of 0.5 microinch are obtainable on hard, dense metal. Finishes of 1 microinch and less are being routinely achieved on plug gauges and valve parts.

A lap finish is unique and cannot be duplicated in any other way. The finish pattern is more multidirectional on parts finished by flat lapping. Any other finish provides directional patterns. Experience indicates that a multidirectional finish has a longer life. Parts which have been ground between dead centers prior to lapping can be flat lapped round to 0.000005". That same part centerless ground could not achieve that accuracy.

OPERATIONAL PROCEDURE

Parts are loaded into the slotted workholder and a small amount of abrasive is applied to the top of the parts or to the bottom of the upper lap. The upper lap is positioned concentrically to the lower lap and then lowered down on top of the parts until the programmed pressure is reached. The machine is started, which rotates the lower lap counterclockwise. This rolls the parts and pushes the workholders around their centers. After the predetermined time (30 seconds), the automatic cycle timer stops the machine. The operator raises the upper lap and swings it out of the way. He checks a few parts for size and rearranges the workload by transposing every part with the opposite one; he also turns the parts end for end as they are transposed. A small amount of lapping oil is added, the top lap is lowered again, and another cycle started.

When the second cycle ends, the upper lap is raised and a few more parts are checked for size. Once again, each part is turned end for end. A small amount of lapping oil is added and the upper lap lowered again. This cycle is repeated until the desired size is reached.

PRODUCTION RATES

These machines can handle about 100 parts per hour, varying with the tolerance required, part size, and machine size. The smaller the part, the more parts can be worked per hour.

Two or three loads per hour can be lapped with an average of 40 parts per load. Of course, this includes handling, checking, and lap conditioning time.

MATERIALS THAT CAN BE LAPPED

Theoretically, any material that can withstand the lapping pressure can be lapped. Hard, stable materials such as glass, hard steels, and certain ceramics are usually lapped with conventional abrasives and cast iron laps. Carbides are lapped with very hard abrasives and a hard, sintered, metal bonded, abrasive lap, otherwise there would be excessive lap wear.

PART CONFIGURATION

Straight, solid pins are the easiest parts to lap. However, any cylindrical parts can be lapped between flat laps. When lapping minor diameters on shafts, the workholder is arranged with the larger diameter overhanging the OD and ID of the laps.

To manage the large variety of jobs, many workholders are available. When working parts to a low microinch finish, the parts should not be allowed to rub against the workholder.

LAPPING ABRASIVES

A variety of abrasives is utilized for cylindrical lapping. For a finish of 1 microinch or better on hard steel, hard alumina, 2–3 micron size, suspended in an oily paste is generally selected. A corundum of 800 grit is often used if much stock is to be removed and finish requirements are broader. Harder materials such as carbides and certain alumina ceramics can be lapped with boron carbides or diamond abrasives.

TYPE OF PARTS

The most common cylindrical parts lapped between flat laps are hydraulic valve spools, plug gauges, piston pins, diesel injector parts, and special armatures.

MACHINE DESIGN

Two annular laps, each mounted on a vertical spindle, are used on cylindrical lapping machines. Depending on the machine model, one or both of the laps rotate. Parts are positioned in a workholder which guides them between the lap faces. The workholder is disc shaped and thinner than the workpieces. It is guided in the center by a pin which can move the workpiece eccentrically to the center of the lower lap.

The workpieces are placed in slots which are tangent (not radial) to the center of the workholder. The rolling action of the parts drives the workholder. In some machines, the lower lap rotates while the upper lap is stationary but self-aligning.

FLAT HONING'S NEW LOOK

This type of honing, around for about 50 years, has been used to obtain flat, parallel surfaces which could not be achieved by grinding alone. Flat surfaces can easily be obtained by lapping or honing in any manner. However, parallelism can be achieved only by using twin lapping or honing machines. When equal amounts of stock are removed from both sides simultaneously, stresses are relieved at the same time that mirror finishes are left on the surfaces. The honing wheel is really a fine grit grinding wheel which runs horizontally at low speeds.

The types of parts suited to this process are those requiring precision surfaces, such as valve plates, ball bearing races, seals, pump components, hardened wear surfaces, and hydraulic valve parts.

ADVANTAGES OF FLAT HONING

- This process produces flat, clean, parallel, reflective surfaces.
- Most materials less than Rc 62 can be flat honed.
- Magnetic or any other type of fixturing is not necessary.

Only the simple holding disc is used, which allows work on nonferrous parts.

- The gentle operation results in little surface damage.
- Stresses are released equally from both sides, which is less likely to cause out-of-flat conditions.
- Finished parts come from this operation clean with only a thin film of mineral oil coating, so elaborate cleaning or rustproofing is not normally required.

WHEELS

In the flat honing process, silicon carbide or aluminum oxide wheels are used. Although the silicon carbide is the more versatile, aluminum oxide is used wherever possible because of its lower cost. The wheels are often a vitrified bond with hardness rated between I and K and a structure of 8 or 9. The grits are generally 150–400. A vitrified wheel is the most economical.

The more expensive resin bonded wheels cut faster because they break down under load faster than vitrified bonded wheels. Because of their polishing effect, they produce a more reflective surface. Resin bonded wheels are more suited to honing harder steels than nonferrous materials.

FLAT, PARALLEL HONING MACHINES

The modern two-wheel honing machine provides fast, consistent production. The honing action does not change all day due to temperature control. Water-cooled wheels, temperature-controlled hydraulics, and cutting fluid aid in producing precision parts. An automatic electronic measuring device controls workpiece thickness to within 0.0002″.

A self-aligning free-floating top wheel provides the ultimate in accuracy without costly, periodic alignment of the spindle. The machine is heavily constructed to minimize deflection. The workholder guides the workpieces in an ever-changing pattern between the revolving wheels to provide even honing action and uniform wheel wear. There is a steady flow of filtered honing oil which cleans, lubricates, and cools the process. A rigid, diamond trueing device reconditions and trues the wheels. The top wheel can be raised and lowered under power, and it can be moved aside for access in loading and unloading at the end of cycles. The wheels and workholders have independent drives which permit reversing direction of rotation.

WORKHOLDING

During the honing process, it is necessary to hold the work-pieces. When honing both sides of the workpieces simultaneously, the workholder guides the parts between the wheels. The workholders must be thinner than the finished pieces. To ensure uniform wheel wear, it is good practice to design the workholders to traverse the parts a little beyond the OD and ID of the hone face. One or two pieces cannot be processed in this manner unless some of the vacant positions are filled with dummy parts.

A popular workholder design uses a set of four or more gear-shaped (sprockets) workholders driven by an inner gear. The workholders rotate inside a stationary ring of pins lining the outside edge of the machine table. This provides a positive drive and traverses the workpieces over the wheel face in a constantly changing pattern which is conducive to uniform wheel wear, thus producing parts of excellent parallelism and uniform size. An independent motor drives the inner ring which permits different speeds for the various job requirements. It also allows periodic reversals of rotation which tend to control wheel flatness.

Numerous holes or cutouts are machined into the workholders to hold the workpieces. There should be sufficient clearance between the parts and their cutouts to permit the workpieces to rotate and align themselves easily during the honing process. The workpieces must not fit tightly in the cutouts.

THE HONING PROCESS

During the honing process, a coolant must be applied for the following reasons.

1. It prevents thermal distortion by cooling the workpieces during honing, and cooling the trueing diamond when the honing wheel is trued up.
2. It prevents wheel face loading by flushing the wheels.
3. It flushes away honed workpiece material and abrasive grains broken off the hone wheel.
4. It lubricates both the workpieces and the wheels, thus reducing friction.

The lubricant should be introduced at the wheel center and allowed to flood over the wheel faces and workpieces by centrifugal action. A thin mineral seal oil seems to work best when flat honing with silicon carbide wheels. Synthetic coolants or soluble oil/water mixtures are less desirable. Insufficient lubrication leads to wheel loading and poor finish.

Filtration down to a maximum of 5 microns in particle size should be continuously conducted. Honing softer materials like iron, aluminum, and soft steel is not generally a problem. When honing harder materials, however, heat develops due to higher pressures and increased friction. Sometimes it is necessary to add a heat exchanger to the system. If heat is allowed to build up, the cutting action slows down.

Hardened workpieces with broad surfaces are more difficult to hone because of the load they present. Hone pressure varies depending on the material and shape. Heavy, solid parts can withstand higher pressures than thin parts. Note that pressures of 5–10 psi are common. While too little pressure is usually harmless except for lengthening the job, too much pressure will cause excessive heat, dulling of the grains, and glazing of the surface being honed.

WHEEL TRUEING

The two faces of the annular-shaped bonded abrasive honing wheels must be absolutely flat in order to hone flat-to-close tolerances. This is usually accomplished by using a heavy pivoting arm to traverse a diamond tool across the faces of the two wheels while they are rotating.

The hone wheel faces must be smooth to produce a low microinch finish. This is done by using a slow traverse speed when trueing with a single face diamond. If a diamond cluster is used, the traverse speed can be increased considerably.

When honing harder materials with large surface areas, the honing action can be improved by trueing the wheels with a single point diamond at a rate of 0.06″ per revolution. There are methods to extend the time between wheel trueups. For example, when the honing action starts to slow down, the operator can shut off the coolant while the honing continues. The wheel face will then break down and make the worn, dull grains fall out and expose new sharp grains. This can be done only once between conventional trueings.

GENERAL PRINCIPLES

Parts are presented to the flat honing machine in such a condition as to do as little honing as possible. The principles of flat honing are as follows.

1. Usually, two flat honing wheels are used.
2. Hone wheel faces are maintained in precision flatness.
3. Workpieces are guided in workholders between flat hone surfaces.
4. Workpieces are always free to align themselves during honing; they are never clamped, held rigid, or tightly fixtured.
5. Usually, a light oil (coolant) is used to lubricate and cool as well as keep the hone cutting surface clean.
6. The hones apply pressure to the workpieces to provide the proper cutting action.

CONDITIONS

Flat honing is designed to correct out-of-flatness and out-of-parallelism, as well as improve the surface finish and produce a desired thickness. Typical surface conditions prior to flat honing are: over size, out-of-parallel, bowed, and rough. Usually a lapped or honed surface is a baseline from which important dimensions are taken; there are places where 0.0001″ out-of-flatness or out-of-parallelism could ruin an assembly, for example, a hydraulic seal or ball bearing race.

ADVANTAGES OF BATCH MODE PROCESSING

The best possible flatness and parallelism can be obtained by this method for several reasons. Since the wheel faces' flatness are maintained precisely, the surfaces being generated actually mirror this flatness.

If any internal stresses are present in a workpiece, by removing stock from both sides simultaneously, it helps to normalize the part and maintain close tolerances. It is a known fact that if a part has internal stresses, machining one side at a time will create a bow. When working to tolerances of less than 1 micron (0.00004″), such stresses are a factor which must be considered.

Batch mode processing allows a number of workpieces to be honed at the same time, and each piece will be honed to the same thickness. Basically, doing many pieces at one time actually improves uniformity of production.

28

CERAMICS

BACKGROUND

When humans first mastered the use of fire 10,000 years ago, they learned how to make low temperature earthenware in open firing pits. That unsophisticated production was the start of ceramic development. Until 100 years ago, ceramics meant pottery for tableware, clay pipe and brick, and roofing tiles. It was not until about 1820 that silica refractories were first made. About 50 years later, when mass production of iron and steel began in earnest, new refractory ceramics were developed to replace the conventional fireclays. This change was absolutely necessary for the production of iron and steel to reach today's plateau.

During the late 19th century, K. J. Bayer discovered a method of producing aluminum hydroxide by hydrolytic deposition. Since then, there have been no major alterations in Bayer's method used to refine aluminum. Alumina is produced when this aluminum hydroxide is baked in a rotary kiln. The alumina produced this way has a purity of about 99.6%. Purities as high as 99.9% or better can be obtained by an improvement in the process. For applications requiring a high degree of electrical insulation or for extremely high strength, the higher purity material is required.

CLASSIFICATION OF CERAMICS

A scientific dictionary will tell you that a ceramic is any class of inorganic, nonmetallic products which are subjected to a temperature of 1000°F and above during manufacture or use. They include metallic oxides, carbides, nitrides, borides, or any combination or compounds of these.

One possible classification of ceramics is by chemical composition. We have a number of oxides segregated into binary, ternary, and quaternary compounds. Then there are carbides, nitrides, and oxynitrides. It is also possible to classify by minerals, of which there are many.

Another common classification is by molding technique. Ceramics may be molded by hand, mechanical die pressing, isostatic pressing, slip casting, vibratory casting, injection molding, and a number of less popular methods.

ISOSTATIC PRESSING

The simplest method of molding ceramics is die pressing; however, it also has a simple problem: it is extremely difficult to compress the ceramic powder uniformly. This unequal pressure causes strains in the molded parts which affect quality; in addition, any dry pressing creates friction at the walls. To avoid such problems, isostatic pressing has become very popular. By applying pressure from all sides instead of unidirectionally, we have a much more desirable method. Two other names for this process are rubber pressing and hydrostatic molding.

A preliminary molding at low pressure is carried out creating a weak shape. The molded form is then placed inside a thin rubber bag which is subjected to a high hydraulic pressure. This is a very uniform pressure, and the finished parts are more uniform in physical properties.

Sometimes dry powder is poured into a rubber mold which is then inserted into a liquid media, then high pressure is applied through the liquid. Compression occurs equally in all directions. This process eliminates normal wall friction.

HOT ISOSTATIC PRESSING

Hot isostatic pressing densifies the ceramic parts, and closes internal voids, "healing" them by diffusion bonding. This process is explained in Chapter 26.

SLIP CASTING

The ceramic material, in the form of fine powder, is mixed with about 30% of water by volume. The mixture, called a slip, is poured into a mold made of plaster of paris. The plaster of paris absorbs water slowly from the slurry. If the liquid in the slip is a solvent instead of water, the mold is made of a different material to absorb the solvent.

After a predetermined time period, the excess slip is poured from the mold and the part is removed. Control of the casting thickness is determined by the amount of time permitted to pass from pouring the slurry into the mold until pouring off the excess slurry. The detail of the casting is accurate and delicate. One disadvantage of the process is the long drying time it requires.

VIBRATORY CASTING

In this process, ceramic powder is mixed with no more than 10% of water. The mold is vibrated so the powder gets packed very tightly and has a lower shrink rate than slip casting. Since the water content is so much less than in slip casting, the drying time is much less and, in fact, a plaster of paris mold is unnecessary. The detail of parts made by this technique is also accurate and delicate.

DRY PRESSING

In this process, dry granular ceramic powder is compacted in a metal die. The moisture content of the powder varies between 0% and 4%. The free-flowing granules are crushed and densely packed to form a coherent compact. This technique varies slightly as the type of ceramic powder varies. In any case, the process is not good for close dimensional tolerances.

There are several problems peculiar to this process, and cracks are the most serious. A crack is an indication of a malfunction, and must be dealt with immediately. Die wear is the first thing to check. Cracks can also be caused by entrapped air, bent die pins, excessive friction between die and the powder, and improper ejection from the die. A compact will expand about 0.5% immediately after ejection, which can cause cracks.

Unless the die is made of carbides, many ceramics will cause die wear because they are abrasive. There are so many precautions to be aware of that the best step is to use the services of an experienced die designer.

The dimensional precision of a ceramic part depends on many factors: the particle size and distribution, the green density after compacting, the firing cycle and sintered density, and the inherent shrinkage of the composition. The method which enables ceramics to meet these exacting requirements must also be economical and suitable for large-scale production. Steel die pressing comes closest to fulfilling these requirements and, accordingly, has been developed to a high degree of automation.

SELECTING THE POWDERED MATERIAL

Ceramic parts are made of main constituents and auxiliary constituents (which help the sintering of the main). The two constituents must be evenly mixed for best results. Mechanical pulverization followed by classification is the most widespread method of preparing ceramic powder. Two significant determinants of the sintering qualities of parts are powder size and grain distribution.

To further increase the performance of ceramic products, ultrafine powders must be made purer. Materials with these qualities are being developed by using several modifications to the basic process.

High purity alumina is a proven tough performer in corrosive, heat, and wear environments. It is an excellent electrical insulator and is displacing other materials in rugged applications. Alumina is as hard as sapphire and outperforms some of the hardest known materials in situations calling for extreme hardness. See Fig. 28.1.

Fig. 28.1. Alumina products. (Courtesy of WESGO Division, GTE.)

The compressive strength of alumina exceeds that of glass, porcelain, or quartz. At elevated temperatures, up to 2000°F, alumina is dimensionally stable under heavy load, while metals would warp and flow plastically. Alumina has a fairly high thermal conductivity compared to other ceramics, being roughly equivalent to Fe-Ni or Fe-Cr alloys. This material is inert to oxidation and is virtually unaffected by water, solvents, and salt solutions.

Most companies manufacturing ceramic powders offer different categories of powder purity ranging from 94 to 99.5% pure alumina to cover a wide range of applications. As purity increases, properties such as chemical resistance, thermal conductivity, dielectric strength, and resistivity increase. But higher purity also means higher cost.

TOLERANCES

Both microminiature parts and cylinders 20″ in diameter and 5' long can be pressed out of ceramic powder; and there are companies which can already handle even larger parts. The tolerances are a function of size. Normal as-fired tolerances are ±1% or ±0.005″ on all dimensions, whichever is greater. Depending on the size and shape of the part, standard grind and lap tolerances achievable are as follows.

Grind Tolerances	± 0.0002″ on all dimensions
	Finish 16 microinches
	Flatness 0.001 in/in
Parallelism	To 0.00002″
Flat Lapping	0.000012″
	3 microinch is attainable.

SILICON NITRIDE

One of the toughest ceramics known to man is silicon nitride; and it is still considered an advanced engineering material. Silicon nitride is already meeting industry's demand for performance at levels other conventional metals and ceramics cannot meet. It has low density (less than ½ that of stainless steel).

Fig. 28.2. Silicon nitride products. (Courtesy of WESGO
Division, GTE.)

Above 1000°C, it has a bending strength above 100,000 psi.
Besides excellent corrosion resistance, it has low thermal expansion, low thermal conductivity, a high thermal shock resistance,
and fracture toughness. See Fig. 28.2.

Several applications include pistons, turbine blades, nozzles,
bushings, bearings, and gears. Lately, other applications include
missile radomes, turbine shrouds, electronic substrates, and
pump and valve parts for the chemical industry. Many parts
using this material are still in development, so they are made in
small quantities. Yet there are some, already doing good work
in industry, which are made in large quantities.

WESGO has made a tailpipe and combustion chamber for

a pulse combustion engine out of silicon nitride. This may be the world's largest component made from this material. The two parts weigh about 70 lbs each, but most of the parts made are much smaller than these. See Fig. 28.3.

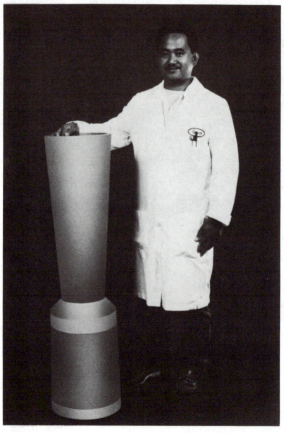

Fig. 28.3. Tailpipe and combustion chamber for a pulse combustion engine made from silicon nitride. (Courtesy of WESGO Division, GTE.)

BRAZING

The best brazing alloys are spherical in shape, low in gas content, and free of impurities. They are available as ribbons, sheets, rings, washers, special preforms, powders, and pastes. They are used in vacuum, hydrogen, or inert gas furnaces. Certain brazing alloys can meet industry's requirements for high strength, hermeticity, and reliability in ceramic-to-ceramic or ceramic-to-metal bonds without prior metallization. Depending upon the contents of the brazing material, the alloy system will allow brazing at temperatures of 1380, 1545, or 2050°F.

Any process must be commercially successful before it can attain popularity. In the near future, it will be necessary for advanced ceramics to permit successful and economical brazing of ceramics to steel parts. For ceramic engine parts, joints must be made between ceramics and iron and ceramics and steel. In the electronic vacuum industry, these joints have already been made for many years.

Joints can be made using epoxy adhesives, mechanical fasteners, or low melting glass frits. The method selected must survive specific conditions such as thermal cycling, high temperatures for extended periods, vibrations, corrosive conditions, or high stresses. The major problem with using ceramics at all is their brittleness or low tensile strength. Consequently, machined surface defects, such as microfractures caused by grinding, must be removed by lapping or honing. In some cases, alumina can be refired to heal the cracks. Of course, the surfaces would have to be used without further machining.

Currently, the conventional practice of joining metal to ceramics involves a two-step molybdenum metallizing process requiring both nickel plating and brazing. This is expensive and difficult to control. WESGO has developed a one-step process using a silver–copper brazing alloy containing 2% titanium. This alloy permits vacuum brazing in one step. Lower melting point brazing alloys have also been developed which permit successive joints to be brazed near each other when the subsequent brazes are made with lower melting alloys.

In these situations, several criteria must be met. The filler metal must have suitable ductility, permit controlled flow to a preselected position, and allow sufficient wetting of both parts. For best results, (blushing) minimum filler flow on both metal and ceramic surfaces is required.

The thermal expansion of both metal and ceramic parts should be uniform from start to finish. Both parts should be heated similarly to about 120°F below the solidus temperature of the filler metal, and should be held there until all parts (including the brazing fixture) reach uniform temperature; then the heat should be increased to about 120°F above the liquidus temperature of the filler metal and maintained for up to 10 minutes before cooling.

A microprocessor-controlled vacuum furnace will minimize thermal stresses on the ceramic. The brazing alloy can sometimes be selected to allow simultaneous brazing and solution heat treatment of the assembly.

The major problem of mixed material brazing is the difference in thermal expansion of the materials. If the difference and the parts are small, a ductile filler is used. If the parts are larger than 5″, a combination of proper design, low brazing temperature, and ductile filler is necessary.

If the thermal expansion difference is large, brazing should be done with a graded seal. A two-step joining technique can be used: a stainless steel–molybdenum braze first, followed by a lower temperature braze to a ceramic ring with a lower temperature brazing alloy. This minimizes joint stresses on cooling to room temperature.

The brazing material available now can easily meet the requirements of aircraft manufacturers and builders of high power vacuum tubes. The filler material is oxide free and has low vapor pressure. Advanced brazing techniques can provide sophisticated, complex subassemblies. The information on brazing given here merely skims the surface; the material manufacturers should be contacted directly for esoteric information.

REFRACTORIES

There are many factors which must be considered when selecting refractory materials including ease of installation, insulating characteristics, maintenance costs, and operating conditions. The furnace load can vary in one day from 300 to 3000 lbs; the temperature can vary from 1200 to 2400°F. This constant thermal shock can create extensive refractory damage.

Some refractories retain heat so long, a 16-hour cool-down period may be required before repairs can be made. To alleviate problems such as these, a ceramic fiber lining was tried and found to be successful. The bottom line was cost and efficiency, and the ceramic liners are now being used extensively.

THE FIRING PROCESS

After the parts have been pressed — by whatever process available — they are in the "green" state, which means they are weak and easily susceptible to breakage. In this state, they can be machined by conventional means such as milling, turning, or drilling. Since the vendor knows how much the parts will shrink in sintering (the percentage of binder in the powder mix), he can machine very close to final size, leaving just a little for the finish.

Next, the parts are fired and hardened into a new state. This process is called sintering and is similar to the sintering process used in powder metallurgy. The sintering takes place in either batch or continuous kilns. In batch kilns, the parts to be fired are loaded onto ceramic shelves, placed in the kilns, fired for a set period, cooled, and removed. The batch kiln adjusts to firing conditions and is relatively inexpensive. The continuous kiln is generally used for high-volume production. It has three zones for preheating, firing, and cooling. The parts are loaded onto carts or ceramic carriers which enter the kiln at one end, one at a time, at predetermined time intervals. The cart advances on a conveyor through the three zones until the parts are sintered.

Parts without tight tolerances just get cleaned up for delivery. Parts with tight tolerances are ground to the finish dimensions. If they are not severely stressed in use, they will be delivered. However, if they function under high stress, they will next be lapped or honed to get rid of microcracks before delivery. If the part is to be brazed and requires a good finish, it must be microcrack free or the brazing process will enlarge the microcracks and spoil the assembly.

Growing electrical, electronic, aeronautical, atomic energy, and other industries learned that it was necessary to develop ceramics with improved qualities. Development was essential to technical progress; and yet ceramic producers are reluctant to change any detail of their process because of complexities in the relationship between materials and processing methods. Nevertheless, the need for the following parts demanded development: insulators, varistors, capacitors, ferrites (an oxide magnetic material), and biological tools and parts. See Fig. 28.4

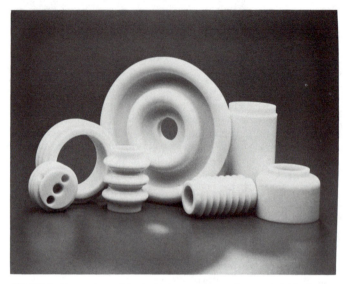

Fig. 28.4. Examples of technical ceramic parts used in industry.
(Courtesy of WESGO Division, GTE.)

MACHINING CERAMICS

After many years of trying various machining methods, procedures evolved which do an excellent job. Vendors know just how much the composition will shrink due to the pressure and then the sintering. This allows them to do most of the machining while the part is in the green condition. The milling, drilling, and turning can be done fairly quickly, although heavy machining forces should be avoided. After sintering, when the part has reached full strength, honing, grinding, and lapping are commonly done.

When the grinding process is to be used, there are several things to consider. In addition to the type of grinding wheel, you must decide on the speed of the wheel and the feed. When grinding ceramics, setting the wheel speed higher will produce better cutting and longer wheel life. With any ceramic product, the machining has a large effect on the strength of the part. In fact, when considering the machining process, one should contemplate the effect it will have on productivity.

Then there are a number of more exotic methods of stock removal. Electrolytic grinding is a fast way to remove stock. Ion beam machining is used on piezoelectric elements, quartz oscillators, and glass lenses. Research is being done on the use of electrical discharge machining of certain types of ceramics.

Lasers have been used to machine grooves on ceramics. Lasers also do a good job drilling holes and holding diameters and center distances to close tolerances. Since this is a noncontact machining method, it is not subject to the quality changes caused by heating, tool wear, or pressure. Thus, stable machining is possible on thin delicate parts. Depending on laser wave length, it is possible to machine through a transparent material. Unlike electron beam work, lasers do not require a vacuum in which to operate.

Ultrasonic machines do an excellent job drilling holes on glass, ferrites, and quartz (see Chapter 25). Cut-off methods are numerous. Cutting may be done by ordinary grinding wheels, diamond wheels, abrasive waterjet machines (see Chapter 10), and ultrasonics.

RESEARCH AND DEVELOPMENT

For years, ceramic parts could not be held to close dimensions due to problems with powder purity and with inconsistencies of the binders. In fact, except for flatness and parallelism, which have been held very close for years, most dimensions could not be machined accurately. However, in recent years, fabrication and machining of ceramics have improved considerably. This was mainly because qualities, not available in other materials, became so significant that R&D finally solved the ceramic production problems.

In 1981, Japan's Ministry of International Trade and Industry commissioned private industry to improve the level of ceramic production technology. The actual targets for this research were: producing powder, molding, sintering, and machining. Companies in the United States and Western Europe recognized the importance of this effort and joined the search for answers.

Currently, great interest is being shown by industry in vibratory compacting of both ceramic and metallic powders (remember that vibration agitates the powders so they seek the

Table 28.1. Properties of Ceramics (Courtesy of Wesgo Div., GTE)

Dense Alumina and Properties Property	Unit	Temperature	AL-500	AL-600	AL-300	AL-995
Al$_2$O$_3$ Content	%		94.0	96.0	97.6	99.5
Flexural Strength	psi	Room Temperature (RT)	50,000	53,000	43,000	45,000
	MPa		345	365	296	310
Compressive Strength	psi	RT	>300,000	>300,000	>250,000	>300,000
	MPa		>2070	>2070	>1720	>2070
Density	lbs/in^3	RT	0.132	0.134	0.136	0.139
	g/cc		3.67	3.72	3.76	3.86
Porosity	% water absorption		vacuum tight 0.00	vacuum tight 0.00	vacuum tight 0.00	vacuum tight 0.00
Color	—		white	white	white	white
Hardness	Rockwell 45N		78	79	75	81
Thermal Conductivity	BTU/ft hr °F W/m °K	RT	11.9 20.5	14.8 25.6	15.5 26.8	16.9 29.3
Coefficient of Linear Thermal Expansion	10^{-6}/°C	25°– 200°C	6.3	6.4	6.9	6.9
		200°– 400°C	7.5	7.6	7.8	7.8
		400°– 600°C	8.0	8.2	8.5	8.3
		600°– 800°C	8.6	8.7	8.8	9.0
		800°–1000°C	9.1	9.0	9.0	9.4
	10^{-6}/°F	77°– 390°F	3.5	3.6	3.8	3.8
		390°– 750°F	4.2	4.2	4.3	4.3
		750°–1110°F	4.4	4.6	4.7	4.6
		1110°–1470°F	4.8	4.8	4.9	5.0
		1470°–1830°F	5.1	5.0	5.0	5.2

238

Property	Units	Material 1	Material 2	Material 3	Material 4
Maximum Working Temperature	°C	1600	1620	1650	1725
	°F	2910	2950	3000	3150
Dielectric Strength (0.100" thick under oil)	DC volts/mil (RT)	650	675	1100	800
	DC kilovolts mm	25.6	26.6	43.3	31.5
Te Value	°C	>950	>950	>1000	>975
	°F	>1740	>1740	>1800	>1790
Volume Resistivity	ohm·cm 25°C/77°F	>10^{14}	>10^{14}	>10^{14}	>10^{14}
	300°C/570°F	2.0×10^{12}	2.0×10^{12}	1.0×10^{12}	2.0×10^{11}
	600°C/1110°F	4.6×10^{8}	5.2×10^{8}	2.3×10^{10}	6.0×10^{8}
	900°C/1650°F	3.5×10^{6}	4.1×10^{6}	5.0×10^{8}	2.5×10^{6}

Dielectric Constant (K')

Freq	Material 1 25°C	300°C	500°C	Material 2 25°C	300°C	500°C	Material 3 25°C	300°C	500°C	Material 4 25°C	300°C	500°C
10 MHz	9.07	9.53	9.91	9.30	9.65	10.10	9.53	9.91	10.14	9.58	9.92	10.20
1000 MHz	9.04	–	–	9.20	–	–	9.00	–	–	9.30	–	–
8500 MHz	8.98	9.26	9.40	9.16	9.30	9.45	9.04	9.32	9.54	9.37	9.61	9.82

Dissipation Factor (Tan δ)

Freq	Material 1 25°C	300°C	500°C	Material 2 25°C	300°C	500°C	Material 3 25°C	300°C	500°C	Material 4 25°C	300°C	500°C
10 MHz	0.00026	0.00028	0.00341	0.00030	0.00061	0.00330	0.00004	0.00016	0.00052	0.00003	0.00009	0.00040
1000 MHz	0.00062	–	–	0.00044	–	–	0.00030	–	–	0.00014	–	–
8500 MHz	0.00078	0.00155	0.00155	0.00062	0.00085	0.00121	0.00045	0.00040	0.00072	0.00009	0.00014	0.00025

Loss Factor (K' Tan δ)

Freq	Material 1 25°C	300°C	500°C	Material 2 25°C	300°C	500°C	Material 3 25°C	300°C	500°C	Material 4 25°C	300°C	500°C
10 MHz	0.00236	0.00267	0.03369	0.00279	0.00588	0.03333	0.00038	0.00158	0.00527	0.00029	0.00089	0.0040
1000 MHz	0.00560	–	–	0.00405	–	–	0.00270	–	–	0.00130	–	–
8500 MHz	0.00700	0.01165	0.01457	0.00568	0.00719	0.01143	0.00407	0.00373	0.00687	0.00084	0.000135	0.0024

closest packing possible). Silicon nitride and silicon carbide are both undergoing tests to see if they can perform satisfactorily under load at 2450°F. There are a number of places in heat engines where the automotive and aeronautic industries would like to use these ceramics.

At the University of California, researchers are trying to develop a superior method of manufacturing ceramics. They hope to handle ceramics like pliable plastics instead of solid, powder material. The researchers are now producing silicon nitride coatings in 20 minutes instead of the usual several hours, by using polysilazane precursors at temperatures of 450–800°C instead of the conventional 1700°C. See Table 28.1.

Although there are several American companies performing excellent research on ceramics, there is one in particular, known to the author, which is currently producing small ceramic parts (up to golf ball size) and holding tolerances of ±0.001". These parts require no machining. This chapter should indicate that there is a great future in the ceramic field.